Electron–Lattice Interactions in Semiconductors

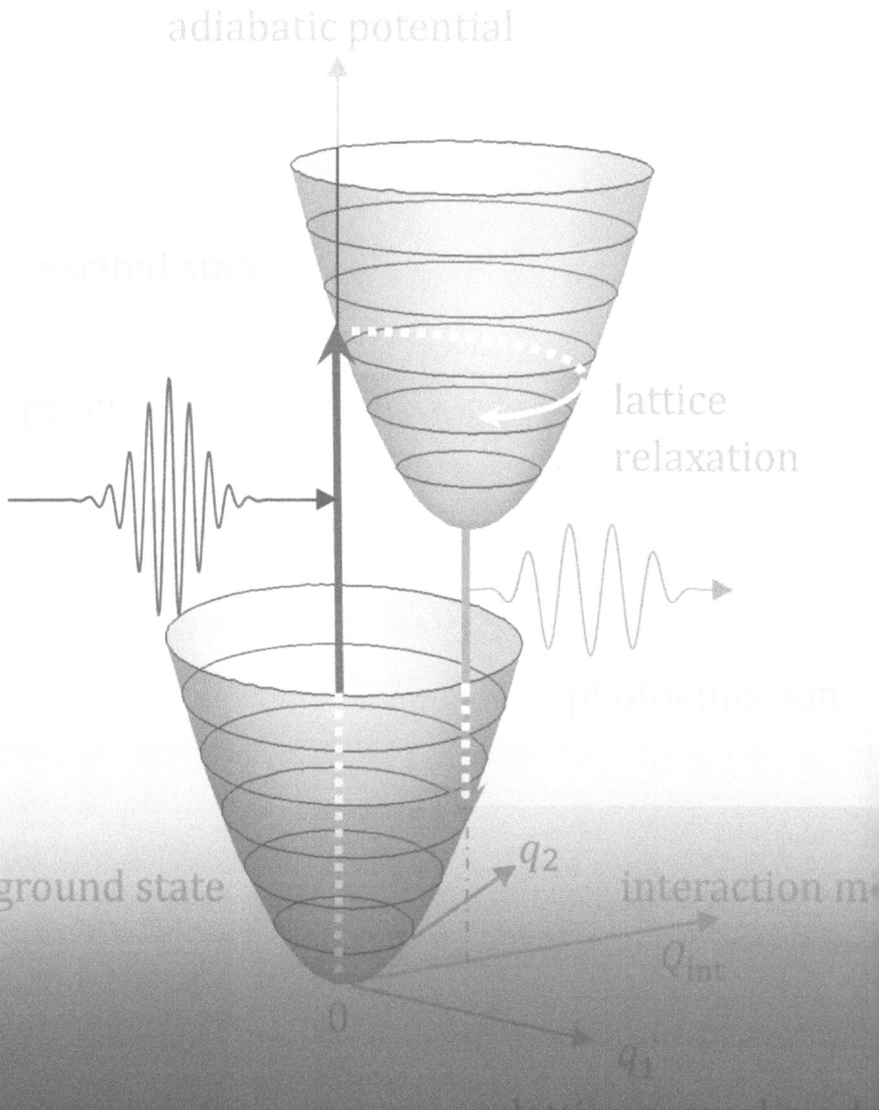

Electron–Lattice Interactions in Semiconductors

Yuzo Shinozuka

JENNY STANFORD
PUBLISHING

Published by

Jenny Stanford Publishing Pte. Ltd.
Level 34, Centennial Tower
3 Temasek Avenue
Singapore 039190

Email: editorial@jennystanford.com
Web: www.jennystanford.com

British Library Cataloguing-in-Publication Data
A catalogue record for this book is available from the British Library.

Electron–Lattice Interactions in Semiconductors

ISBN 978-981-4800-96-9 (Hardcover)
ISBN 978-1-003-02346-3 (eBook)

Contents

Preface

This book is for graduate students who have finished the basic study in solid-state physics and also for research scientists and engineers who participate in materials science and engineering, including the research of electronic and optical semiconductor devices. I did not intend to write a second version of ordinary textbooks on semiconductor physics, but I have presented here underlying physics and theoretical treatments not found in usual textbooks. The necessary knowledge is the basic quantum mechanics and solid-state physics. The target readers are those who are interested in the theoretical approach to semiconductor physics, participating in experimental as well as theoretical researches.

The book starts with a description of the basic properties of condensed matter, treating the system as a macroscopic association of electrons and ions. In their ground state, fruitful results are derived on crystal semiconductors, such as the band theory of electrons in a periodic lattice and the lattice vibrations. The electron–lattice interaction is then introduced as a dynamical response of condensed matter when it is electronically excited. Various optical electro-atomic processes are precisely explained with the aid of proper configuration coordinate diagrams, including carrier scattering, polaron formation, lattice relaxation, Stokes shift and phonon side band in optical spectrum, intrinsic and extrinsic self-trapping, multiphonon process of carrier capture, nonradiative recombination, and symmetry-breaking structural changes.

Half of the contents of this book are somewhat differently described from those of other textbooks on semiconductors. I have tried to show and explain various electronic and atomic processes from a unified point of view. Although the book does not contain many calculation methods to obtain the quantitative value of physical parameters in semiconductors, after reading this book, I hope readers will deeply understand the mechanism of the above-mentioned phenomena, i.e., underlying physics in materials science with fruitful concepts.

This book is dedicated to late Professor Yutaka Toyozawa of the Institute for Solid State Physics, University of Tokyo. He showed me the significance of the optical processes in condensed matter physics and his perspective on the future materials science. Many of the contents in this book are originated from his suggestions and discussion. Special thanks to Professor Keiichiro Nasu and Professor Yosuke Kayanuma for fruitful discussion and sharing their excellent idea in physics since I started my research. I would like to show my greatest appreciation to Professor Koji Maeda and Professor Koichi Murakami for their long-term collaboration in the project "Nano processes by Electronic Excitation" in Japan. Thanks are also due to colleagues and students at Yamaguchi University and Wakayama University for their help, support, and discussion.

I would like to acknowledge all people who have helped me in preparing the manuscript and publishing this book. The support by the JSPS KAKENHI Grant no. JP16H06425 is highly appreciated.

Last but not least, I am very grateful to my wife, Keiko, who assisted me during this time-consuming preparation of the manuscript with great patience and heartful encouragement.

Yuzo Shinozuka
October 7, 2020
Wakayama, Japan

Chapter 1

Perfect Crystal

This chapter is intended as an introduction to the ground state properties of condensed matter that has a crystal lattice structure. Starting from the general basic Hamiltonian of an association of many electrons and ions, we discuss the condensation of atoms. With the resulting crystal lattice structures, we review the basic physical properties of electrons in condensed crystalline materials with particular attention to the translational symmetry, the Bloch theorem, and the band theory. We also discuss the physics of collective atomic motion, i.e., lattice vibrations and their quantum version "phonon" including their thermal properties.

1.1 Basic Hamiltonian

A bulk condensed material consists of a huge number of atoms, i.e., the order of 10^{23}. The motion of an atom in a bulk material is strongly affected by other atoms and also by electrons moving around the atom. The reverse is also true: The motion of an electron in a material is strongly affected by the other electrons and also by the surrounding atoms. In order to discuss the mutual interaction between atoms and electrons in condensed matter, we usually use one of the following three schemes (Fig. 1.1).

Electron–Lattice Interactions in Semiconductors
Yuzo Shinozuka

ISBN 978-981-4800-96-9 (Hardcover), 978-1-003-02346-3 (eBook)
www.jennystanford.com

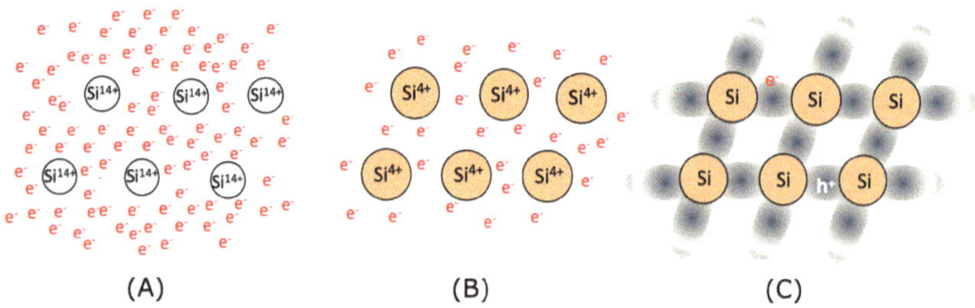

Figure 1.1 Three typical schemes used in condensed matter physics: (A) electrons + nuclei, (B) valence electrons + ions, and (C) several carriers + lattice.

A material is considered to be constructed from

(A) electrons + nuclei,

(B) valence electrons + ions,

(C) several carriers + lattice.

The scheme (C) contains a useful concept of carrier (electron and hole) and lattice vibration, and then is popular in semiconductor physics. However, we choose the scheme (B) in order to clarify the origin of the scheme (C) and its validity [1]. The scheme (A) is necessary when we discuss high-energy phenomena, such as core electron excitations by X-ray irradiation.

In a bulk material, there are many ($N_e \sim 10^{23}$) valence electrons and many ($N_I \sim 10^{23}$) ions, which interact with each other via Coulomb force. The Hamiltonian of the system can be formally written as

$$H = H_e(\boldsymbol{r}_1, \boldsymbol{r}_2, \ldots) + V_{eI}(\boldsymbol{r}_1, \boldsymbol{r}_2, \ldots, \boldsymbol{R}_1, \boldsymbol{R}_2, \ldots) + H_I(\boldsymbol{R}_1, \boldsymbol{R}_2, \ldots),$$

$$H_e(\boldsymbol{r}_1, \boldsymbol{r}_2, \ldots) = K_e + V_e = \sum_i \frac{\boldsymbol{p}_i^2}{2m} + \frac{e^2}{4\pi\varepsilon_0} \sum_i \sum_{i'} \frac{1}{|\boldsymbol{r}_i - \boldsymbol{r}_{i'}|},$$

$$H_I(\boldsymbol{R}_1, \boldsymbol{R}_2, \ldots) = K_I + V_I = \sum_j \frac{\boldsymbol{P}_j^2}{2M_j} + \frac{e^2}{4\pi\varepsilon_0} \sum_j \sum_{j'} \frac{Z_j Z_{j'}}{|\boldsymbol{R}_j - \boldsymbol{R}_{j'}|},$$

$$V_{eI}(\boldsymbol{r}_1, \boldsymbol{r}_2, \ldots, \boldsymbol{R}_1, \boldsymbol{R}_2, \ldots) = -\frac{e^2}{4\pi\varepsilon_0} \sum_i \sum_j \frac{Z_j}{|\boldsymbol{r}_i - \boldsymbol{R}_j|}, \tag{1.1}$$

Here $\boldsymbol{r}_i$ and $\boldsymbol{p}_i$ are, respectively, the position and the momentum of the i-th electron with mass m and charge $-e$; $\boldsymbol{R}_j$ and $\boldsymbol{P}_j$ are, respectively, the position and the momentum of the j-th ion with mass M_j; Z_j stands for the valence number of the j-th ion. We use the MKS units for the Coulomb interaction where ε_0 is the dielectric constant in the vacuum. The Hamiltonian (1.1) can also be used in the scheme (A), where Z_j stands for the atomic number. The interaction terms V_e, V_I, and V_{eI}, respectively, denote the interactions among electron–electron, ion–ion, and electron–ion, all originated from the Coulomb interaction. At present, we will not take care of their detailed forms. With the Hamiltonian (1.1), the Schrödinger equation for valence electrons and ions in the material is formally given by

$$H\Phi_{e,I}(\boldsymbol{r}_1,\boldsymbol{r}_2,\dots,\boldsymbol{R}_1,\boldsymbol{R}_2,\dots) = E\Phi_{e,I}(\boldsymbol{r}_1,\boldsymbol{r}_2,\dots,\boldsymbol{R}_1,\boldsymbol{R}_2,\dots). \quad (1.2)$$

However, it is practically impossible to solve Eq. (1.2) because of huge number of variables which are connected with each other. This difficulty requires us to take the following approximations.

1.2 Adiabatic Approximation

Because the ionic masses M_j($\sim$4.7 × 10^{-26} kg for Si) are much heavier than the electron mass m_e(=9.1 × 10^{-31} kg) the adiabatic (Born–Oppenheimer) approximation can be applied to separate the electronic **fast** motion and the ionic **slow** motion. First, let us fix the ionic positions $\{\boldsymbol{R}\} \equiv (\boldsymbol{R}_1,\boldsymbol{R}_2,\dots)$; then N_e electrons obey the following Schrödinger equation:

$$[H_e(\boldsymbol{r}_1,\boldsymbol{r}_2,\dots) + V_{eI}(\boldsymbol{r}_1,\boldsymbol{r}_2,\dots,\{\boldsymbol{R}\})]\Psi_{\{\boldsymbol{R}\},l}(\boldsymbol{r}_1,\boldsymbol{r}_2,\dots),$$

$$= E_l(\{\boldsymbol{R}\})\Psi_{\{\boldsymbol{R}\},l}(\boldsymbol{r}_1,\boldsymbol{r}_2,\dots). \quad (1.3)$$

Here $\Psi_{\{\boldsymbol{R}\},l}(\boldsymbol{r}_1,\boldsymbol{r}_2,\dots)$ and $E_l(\{\boldsymbol{R}\})$ are the l-th eigenstate and eigenenergy of N_e electrons in the fixed ionic configuration $\{\boldsymbol{R}\}$, which are considered to be parameters for a while. It is still difficult to solve Eq. (1.3) exactly, because the electron–electron interaction term $V_{eI}(\boldsymbol{r}_1,\boldsymbol{r}_2,\dots,\{\boldsymbol{R}\})$ mixes all electronic variables to form many body problem of $3N_e$ freedom. The equation can be reduced to a tractable form by using a proper treatment on many body problems into a single particle problem, such as local density functional (LDF) approximation. We will not go into details on the approximations, but hereafter, we simply assume that we have formally obtained the eigenstates $\Psi_{l,\{R\}}$ and eigenvalues $E_l(\{\boldsymbol{R}\})$ of Eq. (1.3). The adiabatic potential of the ions for the l-th eigenstate of the N_e-electrons is given by

$$U_l(\{\boldsymbol{R}\}) = E_l(\{\boldsymbol{R}\}) + V_I(\{\boldsymbol{R}\}), \tag{1.4}$$

which is a function of $3N_I$ dimensional coordinates $\{\boldsymbol{R}\}$ (Fig. 1.2). The motion of ions, i.e., the wave function χ of ions is considered to obey the following Schrödinger equation:

$$\left[\sum_j \frac{\boldsymbol{P}_j^2}{2M_j} + U_l(\{\boldsymbol{R}\})\right] \chi_{l,n}(\boldsymbol{R}_1, \boldsymbol{R}_2, \ldots) = E_{l,n}\, \chi_{l,n}(\boldsymbol{R}_1, \boldsymbol{R}_2, \ldots). \tag{1.5}$$

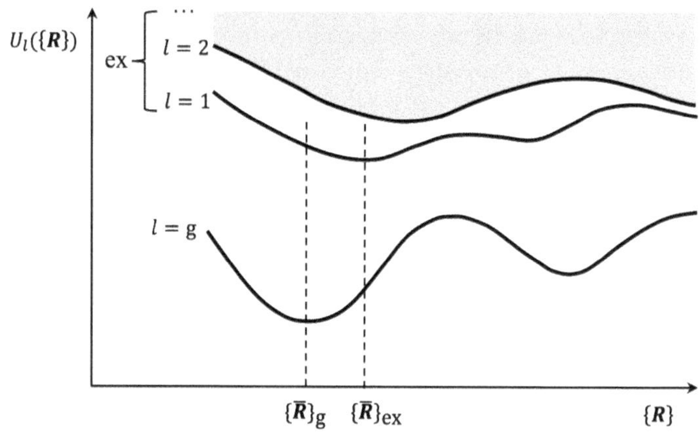

Figure 1.2 Schematic representation of the adiabatic potentials $U_l(\{\boldsymbol{R}\})$ for condensed matter. Note that the ordinate $\{\boldsymbol{R}\} \equiv \boldsymbol{R}_1, \boldsymbol{R}_2, \ldots$ represents $3N_I$ dimensional coordinates.

It should be noted that

(a) The adiabatic potentials $U_l(\{\boldsymbol{R}\})$ reflect all the interactions (V_e, V_I, and V_{eI}) among electrons and ions in the material.

(b) Even along the same l-th branch of $U_l(\{\boldsymbol{R}\})$ the electronic state $\Psi_{\{R\},l}(\boldsymbol{r}_1, \boldsymbol{r}_2, \ldots)$ changes its character continuously with the change of the ionic configuration $\{\boldsymbol{R}\} \equiv (\boldsymbol{R}_1, \boldsymbol{R}_2, \ldots)$.

(c) The direct product $\Psi_{\{R\},l}(\boldsymbol{r}_1, \boldsymbol{r}_2, \ldots)\, \chi_{l,n}(\boldsymbol{R}_1, \boldsymbol{R}_2, \ldots)$ is not a true eigenstate of the original Hamiltonian, $H = H_e + V_{eI} + H_I$, because the momentum $\boldsymbol{P}_j$ in H_I operates on $\Psi_{\{R\},l}$. See Appendix I.

(d) If the ionic variables $\{\boldsymbol{R}\}$ can be treated classical, our alternative task is to solve the set of Newton equations with generalized force $-\partial U_l(\{\boldsymbol{R}\})/\partial \boldsymbol{R}$: This task will be performed in Section 1.5.

1.3 Crystal Structures of Bulk Materials

Before going to solve the Schrödinger Eq. (1.5) with the adiabatic Hamiltonian for ions, we first look at the minimum of the adiabatic potential $U_l(\{\boldsymbol{R}\})$ in Eq. (1.4). For a macroscopic and electrically neutral material, the minimum $\{\bar{\boldsymbol{R}}\}$ of the adiabatic potential $U_g(\{\boldsymbol{R}\})$ for the electronic ground state $\Psi_{\{\bar{\boldsymbol{R}}\},g}(\boldsymbol{r}_1,\boldsymbol{r}_2,...)$ (is considered to) gives a crystal structure that has a translational symmetry. It may happen to give an amorphous structure at a local minimum. (It is not obvious that the lowest energy ionic configuration of a macroscopic homogeneous system always gives a **crystal structure** which has a translational symmetry, but does not give an amorphous phase). The type of the realized crystal structure depends on the constituent of the material. It is the diamond structure for silicon, the zincblende structure for GaAs, ... For carbon, besides the ground graphite structure, there are several metastable structures: diamond, C_{60}(fullerene), nanotube, graphene, ...

Every possible crystal structure can be classified in terms of the translational symmetry and the arrangements of the content ions in the primitive cells (Fig. 1.3). Some typical examples of crystal structures are presented in Figs. 1.4–1.6. In a crystal, it would be physically effective to redefine the stable position $\bar{\boldsymbol{R}}_j$ of each ion by a set of the position $\boldsymbol{T}_l$ of the unit cell and the distance $\boldsymbol{s}_j$ from the origin of each cell (Fig. 1.3). Redefine the new index label as $(\boldsymbol{l}, j)$.

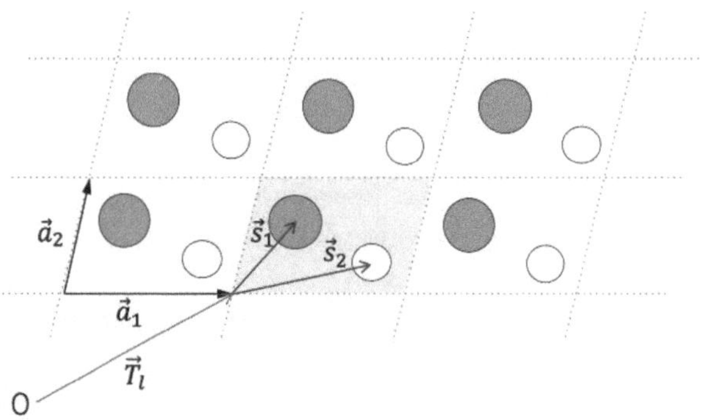

Figure 1.3 Illustration of a typical crystal lattice with a translational symmetry. The constituent ions are located in a unit cell.

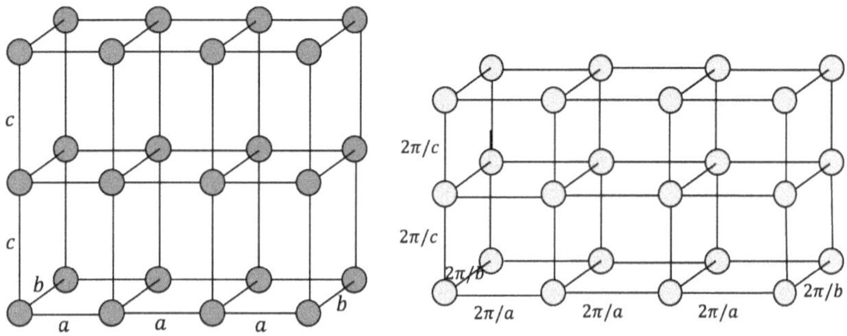

(a) Rectangular *r*-space (b) Rectangular *k*-space

Figure 1.4 Unit cells in real space and reciprocal space for a rectangular lattice.

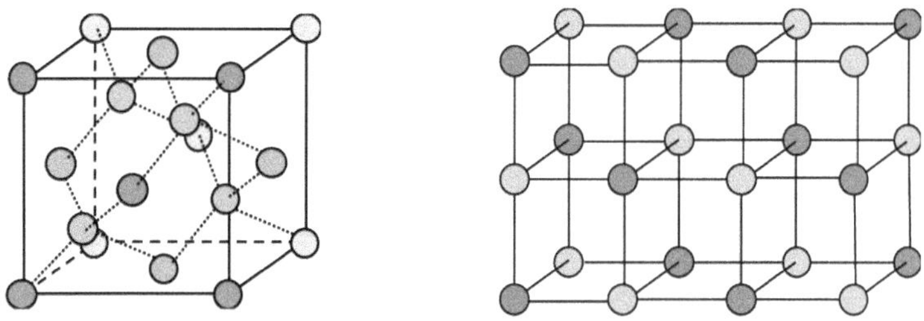

(a) zincblende structure (b) sodium chloride structure

Figure 1.5 Typical crystal structures.

$$\bar{\boldsymbol{R}}_{l,j} = \boldsymbol{T}_l + \boldsymbol{s}_j,$$

$$\boldsymbol{T}_l = l_1\boldsymbol{a}_1 + l_2\boldsymbol{a}_2 + l_3\boldsymbol{a}_3, \qquad \boldsymbol{l} = (l_1, l_2, l_3) \quad \text{integer},$$

$$\boldsymbol{s}_j = s_{j1}\boldsymbol{a}_1 + s_{j2}\boldsymbol{a}_2 + s_{j3}\boldsymbol{a}_3, \quad j = 1, 2, \ldots, S. \;\; 0 \le s_{j1}, s_{j2}, s_{j3} \le 1 \tag{1.6}$$

Here $\boldsymbol{a}_1$, $\boldsymbol{a}_2$, $\boldsymbol{a}_3$ are the primitive translational vectors. The total number of ions are $N \times S$, where N is the number of unit cells each contains S ions.

The reciprocal primitive lattice vectors $\boldsymbol{b}_1$, $\boldsymbol{b}_2$, $\boldsymbol{b}_3$ are defined as

$$\boldsymbol{b}_1 = \frac{2\pi\, \boldsymbol{a}_2 \times \boldsymbol{a}_3}{\boldsymbol{a}_1 \cdot (\boldsymbol{a}_2 \times \boldsymbol{a}_3)}, \; \boldsymbol{b}_2 = \frac{2\pi\, \boldsymbol{a}_3 \times \boldsymbol{a}_1}{\boldsymbol{a}_2 \cdot (\boldsymbol{a}_3 \times \boldsymbol{a}_1)}, \; \boldsymbol{b}_3 = \frac{2\pi\, \boldsymbol{a}_1 \times \boldsymbol{a}_2}{\boldsymbol{a}_3 \cdot (\boldsymbol{a}_1 \times \boldsymbol{a}_2)}, \tag{1.7}$$

which have properties

$$a_i \cdot b_j = 2\pi\delta_{ij}. \tag{1.7a}$$

The set of $\boldsymbol{b}_1$, $\boldsymbol{b}_2$, $\boldsymbol{b}_3$ spans a unit cell in the reciprocal lattice (**k**-space), so called the Brillouin zone (Fig. 1.6). The reciprocal lattice vectors play important roles in the energy spectrum of electrons and phonons in crystal structures. We will not go into the details here: Readers are advised to look again a list of crystal structures in a suitable textbook on solid-state physics [2–6].

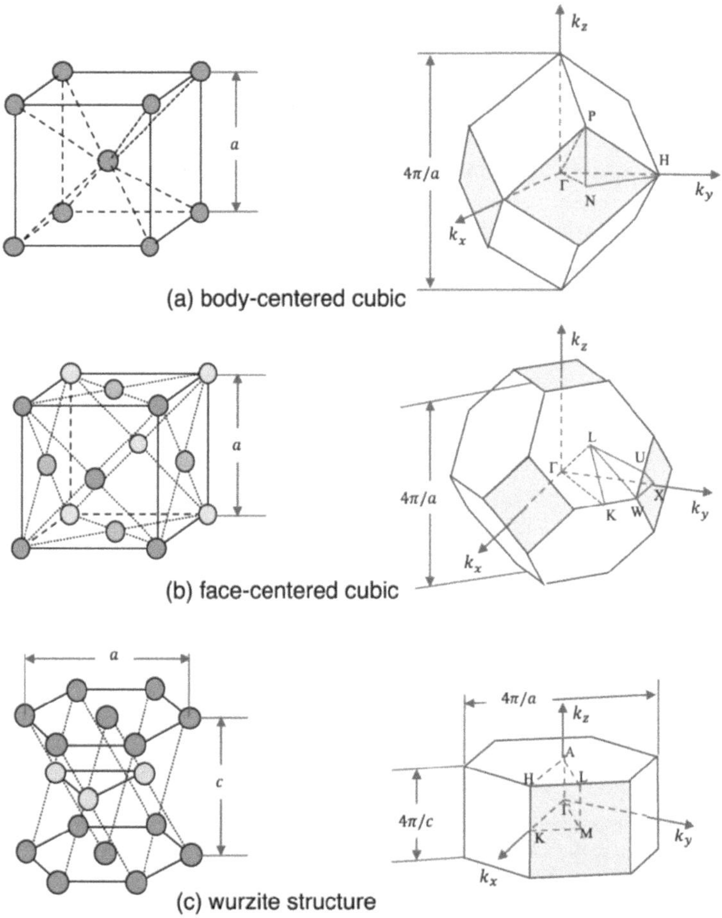

Figure 1.6 Typical crystal structures and the corresponding Brillouin zone.

1.4 Electronic States in Crystals: Band Theory

Let us next discuss electronic states in a material that forms a crystal structure, where each ion locates at their stable position in the periodic lattice $\{\bar{\boldsymbol{R}}\}$. The Hamiltonian of electrons ($i = 1, 2, \ldots, N_\mathrm{e}$) is given by

$$H_\mathrm{e} = \sum_i \left(\frac{\boldsymbol{p}_i^2}{2m} + V(\boldsymbol{r}_i) \right) + \sum_{i,j} V^{\mathrm{e-e}}(\boldsymbol{r}_i, \boldsymbol{r}_j). \tag{1.8}$$

Here $V(\boldsymbol{r}_i)$ is the potential energy felt by an electron at $\boldsymbol{r}_i$ originated from the electron–ion interaction and hence has a translational symmetry. The information of the ionic position $\{\bar{\boldsymbol{R}}\} \equiv \bar{\boldsymbol{R}}_1, \bar{\boldsymbol{R}}_2, \ldots$ has been included implicitly as the crystal structure in Eq. (1.8). We omit the spin freedom of electrons for a moment.

In order to manage the N_e-electron problem (1.8), it is usual to introduce the mean field theory, where the electron–electron interaction term $\Sigma_{j(\neq i)} V^{\mathrm{e-e}}(\boldsymbol{r}_i, \boldsymbol{r}_j)$ is averaged and combined into $V(\boldsymbol{r}_i)$, forming the effective potential with the translational symmetry (plausible assumption)

$$V_\mathrm{eff}(\boldsymbol{r}_i) = V(\boldsymbol{r}_i) + \langle \sum_{j(\neq i)} V^{\mathrm{e-e}}(\boldsymbol{r}_i, \boldsymbol{r}_j) \rangle_\mathrm{AV}, \tag{1.9}$$

felt by each electron. Now the resulting effective Hamiltonian $H_\mathrm{e}^\mathrm{eff}$ is a simple sum of one-electron Hamiltonian as,

$$H_\mathrm{e}^\mathrm{eff} = \langle H_\mathrm{e} \rangle_\mathrm{AV} = \sum_i \left(\frac{\boldsymbol{p}_i^2}{2m} + V_\mathrm{eff}(\boldsymbol{r}_i) \right). \tag{1.10}$$

The problem is reduced to solve the Schrödinger equation for a single electron, which has the same form for each electron. We can omit the suffix i of electrons hereafter, and concentrate on one-electron Hamiltonian,

$$h_\mathrm{e}^\mathrm{eff} \equiv \frac{\boldsymbol{p}^2}{2m} + V_\mathrm{eff}(\boldsymbol{r}). \tag{1.11}$$

Because $V_\mathrm{eff}(\boldsymbol{r})$ has a translational symmetry of the crystal,

$$V_{\text{eff}}(\boldsymbol{r}) = V_{\text{eff}}(\boldsymbol{r} + \boldsymbol{T}_l), \tag{1.12}$$

$$\boldsymbol{T}_l = l_1\boldsymbol{a}_1 + l_2\boldsymbol{a}_2 + l_3\boldsymbol{a}_3, \quad l_1,\ l_2, l_3\text{: integer} \tag{1.12a}$$

then, according to the Bloch's theorem, one-electron eigenstate $\psi_{\sigma,\boldsymbol{k}}(\boldsymbol{r})$ and its energy $\varepsilon_\sigma(\boldsymbol{r})$ are characterized by the band index σ and a wave vector $\boldsymbol{k}$.

$$h_{\text{e}}^{\text{eff}}\,\psi_{\sigma,\boldsymbol{k}}(\boldsymbol{r}) = \varepsilon_\sigma(\boldsymbol{k})\,\psi_{\sigma,\boldsymbol{k}}(\boldsymbol{r}), \tag{1.13}$$

$$\psi_{\sigma,\boldsymbol{k}}(\boldsymbol{r}) = \exp(i\boldsymbol{k}\cdot\boldsymbol{r})\,u_{\sigma,\boldsymbol{k}}(\boldsymbol{r}), \quad u_{\sigma,\boldsymbol{k}}(\boldsymbol{r}) = u_{\sigma,\boldsymbol{k}}(\boldsymbol{r} + \boldsymbol{T}_l). \tag{1.13a}$$

$$\boldsymbol{k} = \frac{n_1}{N_1}\boldsymbol{b}_1 + \frac{n_2}{N_2}\boldsymbol{b}_2 + \frac{n_3}{N_3}\boldsymbol{b}_3, \quad n_1, n_2, n_3\text{: integer} \tag{1.13b}$$

Here, N_i is the number of unit cells along the i-th direction: $N = N_1N_2N_3$ is the total number of unit cells in a crystal. The wave vector $\boldsymbol{k}$'s are uniformly and densely distributed in the reciprocal space: The total number of $\boldsymbol{k}$ in the first Brillouin zone is N. In Eq. (1.13b), $u_{\sigma,\boldsymbol{k}}(\boldsymbol{r})$ is a periodic function of the translational lattice vector $\boldsymbol{T}_l$ given by Eq. (1.12a). The corresponding eigenenergy $\varepsilon_\sigma(\boldsymbol{k})$ of an electron in macroscopic materials is a quasi-continuous function of $\boldsymbol{k}$ in the Brillouin zone and forms a set of energy bands each characterized by the band index σ.

$$\varepsilon_\sigma(\boldsymbol{k}) = \varepsilon_\sigma(\boldsymbol{k} + \boldsymbol{G}_m), \tag{1.14}$$

$$\boldsymbol{G}_m = m_1\boldsymbol{b}_1 + m_2\boldsymbol{b}_2 + m_3\boldsymbol{b}_3, \quad m_1, m_2, m_3\text{: integer}$$

Here $\boldsymbol{G}_m$ is a reciprocal lattice vector spanned by the reciprocal primitive lattice vectors $\boldsymbol{b}_1$, $\boldsymbol{b}_2$, $\boldsymbol{b}_3$ given in Eq. (1.7). Figure 1.7 shows an example of the electron energy band spectrum in Si and GaAs.

It is useful to introduce the energy density of states defined by

$$D(\varepsilon) = \sum_{\sigma,\boldsymbol{k}} \delta\big(\varepsilon - \varepsilon_\sigma(\boldsymbol{k})\big). \tag{1.15}$$

The energy bands of an electron in semiconductors and insulators are classified into the valence bands (VB) and the conduction bands (CB), which are separated by an energy gap E_g: In the ground state the VB is completely filled with valence electrons while those in CB are completely empty (Fig. 1.7). Hereafter, we will concentrate on the highest valence band. Extension to multiple valence bands is straightforward.

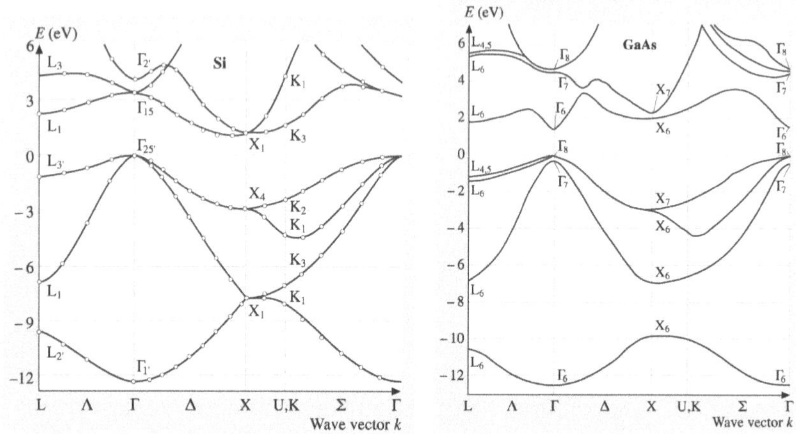

Figure 1.7 The electron energy band structure in Si and GaAs. Reproduced with permission from Ref. [7].

In the ground state $\Psi^{N_e}_{g,\{\bar{R}\}}$ of N_e electrons in a crystal, the lowest-energy one-electron states are occupied according to the Pauli exclusion principle. In semiconductors and insulators, the ground state of N_e electrons can be expressed by an $N_e \times N_e$ Slater determinant composed of one-electron eigenstates $\psi_{v,\boldsymbol{k}}(\boldsymbol{r})$ in the valence band.

$$\Psi^{N_e}_{\{\bar{R}\},g} = \frac{1}{\sqrt{N_e!}} \begin{vmatrix} \psi_{v,\boldsymbol{k}_1}(\boldsymbol{r}_1) & \psi_{v,\boldsymbol{k}_1}(\boldsymbol{r}_2) & \psi_{v,\boldsymbol{k}_1}(\boldsymbol{r}_3) & \cdots \\ \psi_{v,\boldsymbol{k}_2}(\boldsymbol{r}_1) & \psi_{v,\boldsymbol{k}_2}(\boldsymbol{r}_2) & \psi_{v,\boldsymbol{k}_2}(\boldsymbol{r}_3) & \cdots \\ \psi_{v,\boldsymbol{k}_3}(\boldsymbol{r}_1) & \psi_{v,\boldsymbol{k}_3}(\boldsymbol{r}_2) & \psi_{v,\boldsymbol{k}_3}(\boldsymbol{r}_3) & \cdots \\ \cdots & \cdots & \cdots & \cdots \end{vmatrix}. \tag{1.16}$$

It is worthy to remind here that this ground state can be also described in terms of the Wannier orbitals for one-electron state (Fig. 1.8),

$$\phi_{v,\boldsymbol{T}_l}(\boldsymbol{r}) = \frac{1}{\sqrt{N}} \sum_{\boldsymbol{k}} \exp(-i\boldsymbol{k} \cdot \boldsymbol{T}_l)\, \psi_{v,\boldsymbol{k}}(\boldsymbol{r}), \tag{1.17}$$

as

$$\Psi^{N_e}_{\{\bar{R}\},g} = \frac{1}{\sqrt{N_e!}} \begin{vmatrix} \phi_{v,T_1}(r_1) & \phi_{v,T_1}(r_2) & \phi_{v,T_1}(r_3) & \cdots \\ \phi_{v,T_2}(r_1) & \phi_{v,T_2}(r_2) & \phi_{v,T_2}(r_3) & \cdots \\ \phi_{v,T_3}(r_1) & \phi_{v,T_3}(r_2) & \phi_{v,T_3}(r_3) & \cdots \\ \cdots & \cdots & \cdots & \cdots \end{vmatrix}. \tag{1.18}$$

This is because a set of Bloch states and a set of Wannier states are transformed each other by a unitary transformation. In other words, in the ground state of nonmetallic materials, all the Bloch states, $\psi_{v,\boldsymbol{k}}(\boldsymbol{r})$ defined by the $\boldsymbol{k}$ vector, in the valence band are fully occupied, and it can also be expressed that all the Wannier states, $\phi_{v,T_l}(\boldsymbol{r})$ located at each unit cell $\boldsymbol{T}_l$, are fully occupied.

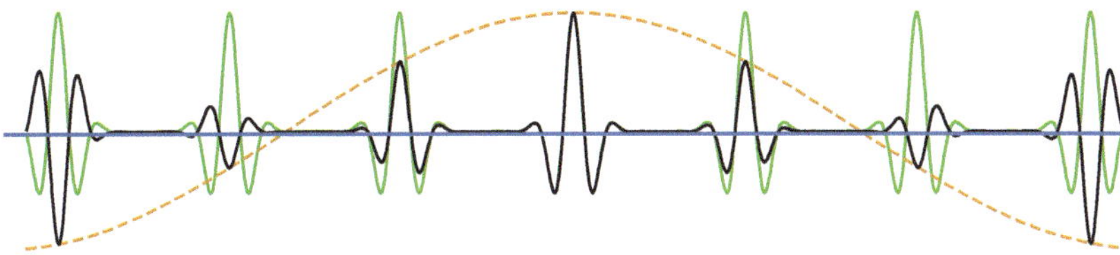

Figure 1.8 Typical profile of a Bloch state (black) and a set of Wannier states (green).

Before closing this section, it would be worthwhile to introduce the second quantized form of many-electron system. Let define the creation operator $a^+_{\sigma,\boldsymbol{k}}$ that operates the electron-vacuum state, $|0\rangle$, to create an electron at the Bloch state $\psi_{\sigma,\boldsymbol{k}}(\boldsymbol{r})$ in the σ band.

$$a^+_{\sigma,\boldsymbol{k}}|0\rangle \quad \leftrightarrow \quad \text{an electron occupies } \psi_{\sigma,\boldsymbol{k}}(\boldsymbol{r}). \tag{1.19}$$

Also define the annihilation operator $a_{\sigma,\boldsymbol{k}}$ which destroy an electron at $\psi_{\sigma,\boldsymbol{k}}(\boldsymbol{r})$. Those operators obey the following anticommutation relations for Fermi particles.

$$\{a_{\sigma,\boldsymbol{k}}, a^+_{\sigma',\boldsymbol{k}'}\} \equiv a_{\sigma,\boldsymbol{k}}\, a^+_{\sigma',\boldsymbol{k}'} + a^+_{\sigma',\boldsymbol{k}'} a_{\sigma,\boldsymbol{k}} = \delta_{\sigma\sigma'}\, \delta_{\boldsymbol{k}\boldsymbol{k}'},$$

$$\{a_{\sigma,\boldsymbol{k}}, a_{\sigma',\boldsymbol{k}'}\} \equiv a_{\sigma,\boldsymbol{k}}\, a_{\sigma',\boldsymbol{k}'} + a_{\sigma',\boldsymbol{k}'} a_{\sigma,\boldsymbol{k}} = 0,$$

$$\{a^+_{\sigma,\boldsymbol{k}}, a^+_{\sigma',\boldsymbol{k}'}\} \equiv a^+_{\sigma,\boldsymbol{k}}\, a^+_{\sigma',\boldsymbol{k}'} + a^+_{\sigma',\boldsymbol{k}'} a^+_{\sigma,\boldsymbol{k}} = 0. \tag{1.20}$$

The ground state $\Psi^{N_e}_{\{\bar{R}\},g}$ of N_e electrons in a crystal (1.16) can be expressed as a multiple of many creation operators for Bloch electrons in the valence band operating to the vacuum $|0\rangle$:

$$\left|\Psi^{N_e}_{\{\bar{R}\},g}\right\rangle = \prod_{k=\text{occupied}} a^+_{v,k}|0\rangle = \prod_{T_l} a^+_{v,T_l}|0\rangle \tag{1.21}$$

Here the vacuum $|0\rangle$ is defined as the completely empty state: No valence electron. As stated above, $\Psi^{N_e}_{\{\bar{R}\},g}$ can be also expressed with a^+_{v,T_l}, the electron creation operator at the Wannier orbital at a unit cell T_l, which is related with $a^+_{v,k}$ under a Fourier transformation: $(\sigma = c, v)$

$$a^+_{\sigma,T_l} = \frac{1}{\sqrt{N}}\sum_{k} \exp(i\boldsymbol{k}\cdot\boldsymbol{T}_l)\, a^+_{\sigma,k}, \tag{1.22a}$$

$$a_{\sigma,T_l} = \frac{1}{\sqrt{N}}\sum_{k} \exp(-i\boldsymbol{k}\cdot\boldsymbol{T}_l)\, a_{\sigma,k}. \tag{1.22b}$$

Up to here, we have not considered the spin freedom. It can be easily included the spin variable $\alpha = \uparrow, \downarrow$ in the above discussion and all the relevant equations by introducing another suffix α to physical quantities.

1.5 Lattice Vibrations in Crystals

1.5.1 Lagrangian of the Lattice

Let us move to the motion of ions around the minimum point $\{\bar{R}\}_g$ of the adiabatic potential $U_g(\{R\})$, where all electrons are in the ground state $\Psi^{N_e}_{\{\bar{R}\},g}$: The valence band states are completely filled with electrons while the conduction band states are empty. Each ion can move around each equilibrium position $\bar{\boldsymbol{R}}_{l,j} = \boldsymbol{T}_l + \boldsymbol{s}_j$, where the index label (l, j) is defined by Eq. (1.6). Let us define a new variable for the ionic displacement (Fig. 1.9)

$$\Delta\boldsymbol{R}_{l,j} \equiv \boldsymbol{R}_{l,j} - \bar{\boldsymbol{R}}_{l,j}. \tag{1.23}$$

Then the Lagrangian of the ion system will be expressed as

$$L = \frac{1}{2}\sum_{l,j} M_j \Delta \dot{\boldsymbol{R}}_{l,j}{}^2 - U_{\mathrm{g}}(\{\Delta \boldsymbol{R}\}). \tag{1.24}$$

Here, $\dot{\boldsymbol{R}}$ denotes the time derivative of $\boldsymbol{R}$ and $\{\Delta \boldsymbol{R}\} = (\Delta \boldsymbol{R}_{l1,1}, \Delta \boldsymbol{R}_{l1,2}, \ldots, \Delta \boldsymbol{R}_{l2,1}, \Delta \boldsymbol{R}_{l2,2}, \ldots\})$. As far as the displacement $\Delta \boldsymbol{R}_{l,j}$'s are small, i.e., much smaller than the lattice constant d, one can formally expand the adiabatic potential $U_{\mathrm{g}}(\{\Delta \boldsymbol{R}\})$ in power series.

$$U_{\mathrm{g}}(\{\Delta \boldsymbol{R}\}) = U_{\mathrm{g}}(\{\Delta \boldsymbol{R} = \boldsymbol{0}\}) + \sum_{l,j,\alpha} \left.\frac{\partial U_{\mathrm{g}}(\{\Delta \boldsymbol{R}\})}{\partial \Delta R^{\alpha}_{l,j}}\right|_{\Delta \boldsymbol{R}=\boldsymbol{0}} \Delta R^{\alpha}_{l,j}$$
$$+\frac{1}{2}\sum_{l,j,\alpha}\sum_{l',j',\beta} \left.\frac{\partial^2 U_{\mathrm{g}}(\{\Delta \boldsymbol{R}\})}{\partial \Delta R^{\alpha}_{l,j}\partial \Delta R^{\beta}_{l',j'}}\right|_{\Delta \boldsymbol{R}=\boldsymbol{0}} \Delta R^{\alpha}_{l,j}\Delta R^{\beta}_{l',j'} + \cdots. \tag{1.25}$$

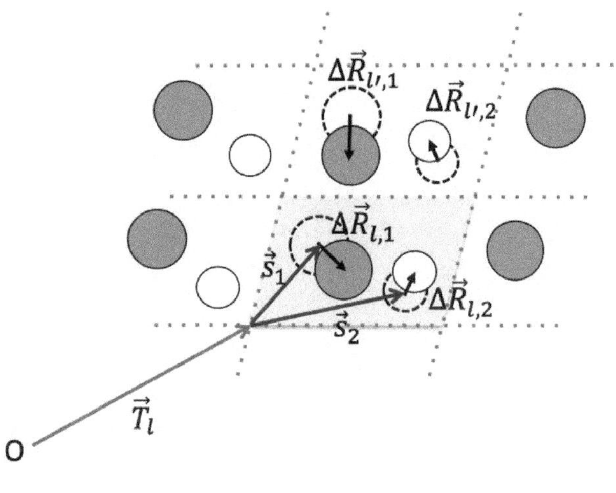

Figure 1.9 The displacement $\Delta \boldsymbol{R}_{l,j}$ of the ion at $\overline{\boldsymbol{R}}_{l,j} = \boldsymbol{T}_l + \boldsymbol{s}_j$. Dotted circle denotes the stable (=home) position of each ion.

Here $\Delta R^{\alpha}_{l,j}$ denotes the $\alpha(=x,y,z)$ component of the displacement $\Delta \boldsymbol{R}_{l,j}$ of j-th ion (at $\boldsymbol{s}_j$) in the unit cell at $\boldsymbol{T}_l$. The total number of these variables are $3NS$. In the expansion (1.25), the origin of the potential energy can be set as $U_{\mathrm{g}}(\{\Delta \boldsymbol{R} = \boldsymbol{0}\}) = 0$ without loss of generality. The coefficients of the first derivative are all zero at the stable position, then the second term in Eq. (1.25) is zero. Thus, the expansion of the adiabatic potential $U_g(\{\Delta \boldsymbol{R}\})$ starts with the quadratic terms whose coefficient

$$\Xi^{\alpha,\beta}_{l,j;l',j'} \equiv \left. \frac{\partial^2 U_g(\{\Delta \boldsymbol{R}\})}{\partial \Delta R^{\alpha}_{l,j} \partial \Delta R^{\beta}_{l',j'}} \right|_{\Delta \boldsymbol{R}=0} \tag{1.26}$$

is called the force coefficient and stands for the mutual interaction between the ion at $\bar{\boldsymbol{R}}_{l,j} = \boldsymbol{T}_l + \boldsymbol{s}_j$ and the ion at $\bar{\boldsymbol{R}}_{l',j'} = \boldsymbol{T}_{l'} + \boldsymbol{s}_{j'}$ mediated by valence electrons around them. In principle, the coefficients can be calculated theoretically: Details will be discussed in Section 1.5.4. At the moment, it is not necessary to have the explicit form of $U_g(\{\Delta \boldsymbol{R}\})$ and hence the values of $\Xi^{\alpha,\beta}_{l,j;l',j'}$: We will proceed a formal discussion.

As a summary, the adiabatic potential for the ionic motion around the crystal position in the harmonic approximation has a quadratic form of $\Delta R^{\alpha}_{l,j}$'s

$$U_g(\{\Delta \boldsymbol{R}\}) = \frac{1}{2} \sum_{l,j,\alpha} \sum_{l',j',\beta} \Xi^{\alpha,\beta}_{l,j;l',j'} \Delta R^{\alpha}_{l,j} \Delta R^{\beta}_{l',j'}. \tag{1.27}$$

In crystals, the force coefficient $\Xi^{\alpha,\beta}_{l,j;l',j'}$ depends only on the relative positions $\boldsymbol{T}_l - \boldsymbol{T}_{l'}$, because of the translational symmetry of the crystal. There are also following relations:

$$\Xi^{\alpha,\beta}_{l,j;l',j'} = \Xi^{\beta,\alpha}_{l',j';l,j} \quad \text{Newton's third law} \tag{1.27a}$$

$$\sum_{l,j} \Xi^{\alpha,\beta}_{l,j;l',j'} = \sum_{l',j'} \Xi^{\beta,\alpha}_{l',j';l,j} = 0 \quad \text{sum rule} \tag{1.27b}$$

The last sum rule (1.27b) holds, because if all atoms displace exactly by the same direction and amount, it basically corresponds to a simple translational shift of the hole crystal, then no ion is affected (no induced force).

1.5.2 Dynamics of Ions [Classical Treatment]

From the Lagrangian (1.24) we can derive the equation for the ionic displacement $\Delta R^{\alpha}_{l,j}$ according to

$$\frac{d}{dt}\left(\frac{\partial L}{\partial \Delta \dot{R}_{l,j}^{\alpha}}\right) - \frac{\partial L}{\partial \Delta R_{l,j}^{\alpha}} = 0, \tag{1.28}$$

which produces

$$M_j \frac{d^2}{dt^2} \Delta R_{l,j}^{\alpha} = -\sum_{l',j'} \Xi_{l,j;l',j'}^{\alpha,\beta} \Delta R_{l',j'}^{\beta}. \quad \alpha, \beta = x, y, z; \; j, j' = 1, \ldots, S \tag{1.29}$$

These are simultaneous differential equations on the dynamical variable $\Delta R_{l,j}^{\alpha}(t)$, whose number is $3NS$. These equations can be solved as follows.

Taking account of the translational symmetry of the crystal and assuming harmonic oscillation both in space with a vector $\boldsymbol{k}$ (defined in Eq. (1.13b)) restricted to the 1st Brillouin zone) and in time with an angular frequency ω, we can set

$$\Delta R_{l,j}^{\alpha} = \frac{1}{\sqrt{M_j}} \sum_{\sigma,\boldsymbol{k}} \eta_{j,\boldsymbol{k}}^{\alpha} \exp\left[i(\boldsymbol{k} \cdot \boldsymbol{T}_l - \omega t)\right], \tag{1.30}$$

then the quantity $\eta_{j,\boldsymbol{k}}^{\alpha}$ obeys,

$$-\omega^2 \eta_{j,\boldsymbol{k}}^{\alpha} = -\sum_{\beta,j'} D^{\alpha,\beta}(j, j'; \boldsymbol{k})\, \eta_{j',\boldsymbol{k}}^{\beta}. \tag{1.31}$$

Here, $D^{\alpha,\beta}(j, j'; \boldsymbol{k})$ is the Fourier transform of the force coefficients $\Xi_{l,j;l',j'}^{\alpha,\beta}$.

$$D^{\alpha,\beta}(j, j'; \boldsymbol{k}) = \frac{1}{\sqrt{M_S M_{S'}}} \sum_{l'} \Xi_{l,j;l',j'}^{\alpha,\beta} \exp[i\boldsymbol{k} \cdot (\boldsymbol{T}_l - \boldsymbol{T}_{l'})]. \tag{1.32}$$

Simultaneous $3S$ Eqs. (1.31) have nonzero solution for $\eta_{j,\boldsymbol{k}}^{\alpha}$'s only if

$$\left| D^{\alpha,\beta}(j, j'; \boldsymbol{k}) - \omega^2 \boldsymbol{I} \right| = 0, \tag{1.33}$$

where $\boldsymbol{I}$ is the unit matrix in $3S \times 3S$ dimension. The matrix $D^{\alpha,\beta}(j, j'; \boldsymbol{k})$ has a relation (* denotes complex conjugate) according to Eq. (1.27a,b),

$$D^{\alpha,\beta}(j,j';-\boldsymbol{k}) = D^{\alpha,\beta}(j,j';\boldsymbol{k})^*, \tag{1.34a}$$

$$D^{\alpha,\beta}(j,j';\boldsymbol{k}) = D^{\beta,\alpha}(j',j;\boldsymbol{k})^*. \tag{1.34b}$$

Since $D^{\alpha,\beta}(j,j';\boldsymbol{k})$ is an Hermitian matrix, we obtain $3S$ real eigenvalues of ω^2 for each wave vector $\boldsymbol{k}$, which can take N discrete values uniformly distributed in the first Brillouin zone. We can set

$$\omega_{\sigma,\boldsymbol{k}} \geq 0 \text{ and } \omega_{\sigma,\boldsymbol{k}} = \omega_{\sigma,-\boldsymbol{k}} \quad \sigma = 1, 2, \ldots, 3S$$

Putting one of the eigen frequency $\omega_{\sigma,\boldsymbol{k}}$ of the solutions of Eq. (1.33) into Eq. (1.31), the ratio $e^{\alpha}_{\sigma,j,\boldsymbol{k}}$ among $\eta^{\alpha}_{j,\boldsymbol{k}}$'s is determined. A set of $(e^{x}_{\sigma,j,\boldsymbol{k}}, e^{y}_{\sigma,j,\boldsymbol{k}}, e^{z}_{\sigma,j,\boldsymbol{k}}) \equiv \boldsymbol{e}_{\sigma,j,\boldsymbol{k}}$ is usually normalized to be a unit vector which characterizes the relative ionic motion in each mode:

$$\text{orthnormality} \sum_{\alpha}\sum_{j} e^{\alpha}_{\sigma,j,\boldsymbol{k}} e^{\alpha}_{\sigma',j,\boldsymbol{k}}{}^{*} = \delta_{\sigma\sigma'}, \tag{1.35a}$$

$$\text{completeness} \sum_{\sigma} e^{\alpha}_{\sigma,j,\boldsymbol{k}} e^{\beta}_{\sigma,j',\boldsymbol{k}}{}^{*} = \delta_{\alpha\beta}\delta_{jj'} \quad \alpha,\beta = x,y,z;\ \ j,j' = 1,\ldots,S. \tag{1.35b}$$

Thus, we obtain a set of normal vibration modes characterized by the mode σ, the wave vector $\boldsymbol{k}$ and the corresponding eigen frequency $\omega_{\sigma,\boldsymbol{k}}$. Typical values of $\omega_{\sigma,\boldsymbol{k}}$ are $\sim 10^{13}$ radHz.

General solutions of the motion of an ion at $\boldsymbol{T}_l + \boldsymbol{s}_j$ are given by a linear combination of the normal modes of (σ, $\boldsymbol{k}$) with each coefficient $f^{\alpha}_{\sigma,\boldsymbol{k}}$

$$\Delta R^{\alpha}_{l,j}(t) = \frac{1}{\sqrt{N\,M_j}} \sum_{\sigma,\boldsymbol{k}} f_{\sigma,\boldsymbol{k}}\, e^{\alpha}_{\sigma,j,\boldsymbol{k}} \exp\left[i\left(\boldsymbol{k}\cdot\boldsymbol{T}_l - \omega_{\sigma,\boldsymbol{k}} t\right)\right], \quad \alpha = x,y,z \tag{1.36}$$

Here N, the total number of unit cells, is introduced so that $f_{\sigma,\boldsymbol{k}}$ is to be a normal coordinate (Section 1.5.6). It should be noted here that Eq. (1.36) does not imply the j-th ion vibrates in proportion to $1/\sqrt{M_j}$ in $(\sigma, \boldsymbol{k})$ mode, because the unit vector $e^{\alpha}_{\sigma,j,\boldsymbol{k}}$ depends on all M_j's in a unit cell.

1.5.3 Lattice Vibration Modes

The normal vibration modes $\{\omega_{\sigma,\boldsymbol{k}}\}$ in crystals are typically classified into the acoustic mode (σ = ac) and the optical modes (σ = op) for materials whose unit cell contains more than two ions. Figure 1.10 schematically shows the acoustic mode and optical mode in a one-dimensional lattice. If a unit cell contains S ions, there are $3S$ vibration normal mode branches. The modes are classified into

3 acoustic modes:

all $\boldsymbol{e}_{ac,j,\boldsymbol{k}}$ are equal within a unit cell

$\omega_{\mathrm{ac},\boldsymbol{k}} \rightarrow 0$ when $\boldsymbol{k} \rightarrow 0$.

$3S$-3 optical modes:

If a unit cell contains two ions A and B,

they move oppositely: $\boldsymbol{e}_{\mathrm{op,A},\boldsymbol{k}} = -\boldsymbol{e}_{\mathrm{op,B},\boldsymbol{k}}$

$\omega_{\mathrm{ac},\boldsymbol{k}} \rightarrow$ finite when $\boldsymbol{k} \rightarrow 0$.

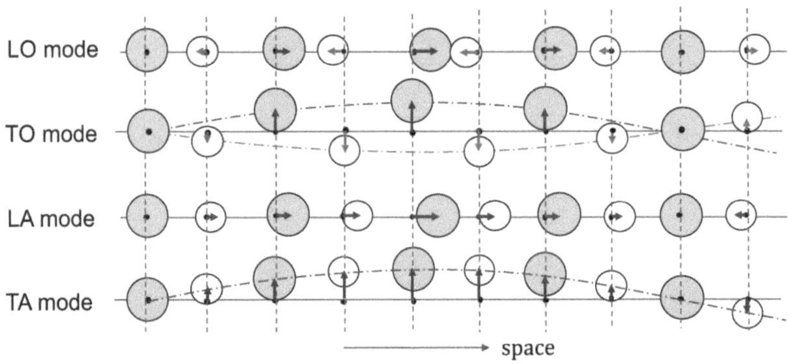

Figure 1.10 Illustration of the optical mode and acoustic mode in a one-dimensional lattice (two ions in a unit cell). L: longitudinal and T: transverse.

In isotropic materials, they are composed of one longitudinal (L) mode $\boldsymbol{e}_{\mathrm{ac},j,\boldsymbol{k}} // \boldsymbol{k}$ and two transverse (T) modes $\boldsymbol{e}_{ac,j,\boldsymbol{k}} \perp \boldsymbol{k}$. Figure 1.11 shows a typical vibrational spectrum of Si and GaAs [8–10].

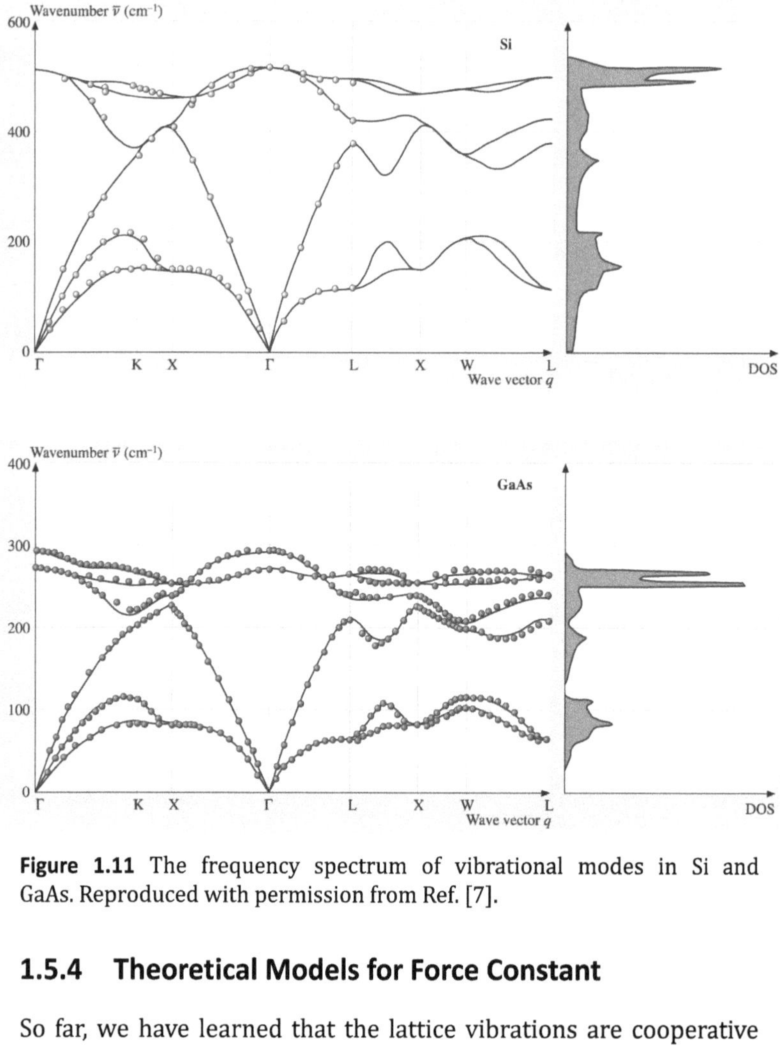

Figure 1.11 The frequency spectrum of vibrational modes in Si and GaAs. Reproduced with permission from Ref. [7].

1.5.4 Theoretical Models for Force Constant

So far, we have learned that the lattice vibrations are cooperative motions of ions (atoms) in a material. In crystal case, they are characterized by the mode σ, the wave vector $\boldsymbol{k}$, and their frequency $\omega_{\sigma,\boldsymbol{k}}$. The eigen frequencies can be calculated in principle if one obtains the information of the force coefficient $\Xi^{\alpha,\beta}_{l,j;l',j'}$ (1.26). There are several models (Fig. 1.12) [4].

(a) Force constant model

This is a generalization of the ball and spring model. Assuming a suitable force constant of the harmonic spring between nearest neighbor ions (atoms), one can obtain the solution of the eigen frequencies, which are fed back to adjust the force constants to fit with experiments. This model works to some extent. Main weakness in this model is that shear stress is not satisfactory taken into account.

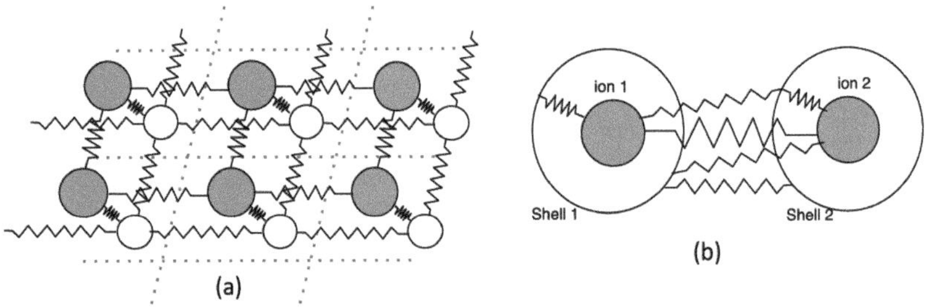

Figure 1.12 (a) Force constant model and (b) shell model, to obtain the force coefficients theoretically.

(b) Shell model

This is an improved version of the Force constant model. One can extend the above model in which each atom is assumed to consist of a rigid ion core surrounded by a shell of valence electrons that can move relative to the cores. This model artificially divides the valence charges between the two atoms involved in the covalent bond. In reality, the valence electrons are "time-shared" between the two atoms in that they all spend part of their time on each atom. The main criticism of the shell model is that the valence electron distributions in the diamond- and zincblende semiconductors are quite different from spherical shells.

(c) Bond model

This model includes the effect of bond bending. These vibrational modes are usually analyzed in terms of valence force fields for stretching the bonds and for changing the angles between bonds (bond bending). The force constants can be determined in a straightforward manner from these valence force fields since the displacements of the ions are related to the bond coordinates. One advantage of this approach is ions are related to the bond coordinates. This pile-up of charge in the covalent bond is known as the bond charge and is well known in the formation of covalent molecules.

(d) Bond Charge model

The force constants can be obtained, in principle, by differentiating the total energy $U_g(\{\Delta \boldsymbol{R}\})$ with respect to the ion coordinates (this is known as the frozen phonon approximation). However, this approach requires intensive computation and the use of supercomputers. The most difficult part of this calculation is to handle the Coulomb interaction between ions and electrons. Owing to the recent rapid progress in the ability of computers, one can directly calculate the adiabatic potential $U_g(\{\Delta \boldsymbol{R}\})$ (1.4) as a function of ionic displacements $\{\Delta \boldsymbol{R}\}$ to some extent. For further details, see Ref. [8].

1.5.5 Note on Electronic States during the Lattice Vibrations

We have found the cooperative motion of ions in a crystal is characterized by a wave vector reflecting the translational symmetry and the mode index depending on the contents of the unit cell. All the interactions among electrons and ions are compiled into the force coefficient $\Phi^{\alpha,\beta}_{l,j;l',j'}$, which is reflected in the phonon modes and their eigen frequencies $\omega_{\sigma,\boldsymbol{k}}$.

One may have a following question on electronic states: since the Bloch states and the electronic ground states $\Psi^{N_e}_{g,\{\bar{R}\}}$ are defined in the rigid crystal lattice where all ions locate strictly at each stable position $\{\bar{\boldsymbol{R}}\}_g$, then once ions are displaced from $\{\bar{\boldsymbol{R}}\}_g$, the electronic states are affected and could be changed from the Bloch states given by Eq. (1.13). This really happens and exactly what we will investigate in the next chapter. However, as far as the total electronic ground state is concerned, it can be treated with a rather simple way, as follows (Fig. 1.13).

In the previous section, we have learned that one-electron eigenstates are the Bloch state $\psi_{\sigma,\boldsymbol{k}}(\boldsymbol{r}) = \exp(i\boldsymbol{k}\cdot\boldsymbol{r})u_{\sigma,\boldsymbol{k}}(\boldsymbol{r})$ (Eq. (1.13)) defined in a crystal lattice where each ion locates strictly at $\{\bar{\boldsymbol{R}}\}_g$, i.e., $\{\Delta \boldsymbol{R}_{l,j} = 0\}$. Let us imagine that there arise displacements of ions $\{\Delta \boldsymbol{R}_{l,j} \neq 0\}$. Then every one-electron eigenstate $\psi_{n,l}(\boldsymbol{r})$

is affected and changed from the Bloch states $\psi_{\sigma,\boldsymbol{k}}(\boldsymbol{r})$. But as long as the displacement $\{\Delta\boldsymbol{R}_{l,j}\}$ is small, a set of electronic state $\psi_{n,l\{\Delta\boldsymbol{R}\}}(\boldsymbol{r})$'s in $\{\Delta\boldsymbol{R}_{l,j} \neq 0\}$ is constructed from a linear combination of $\psi_{\sigma,\boldsymbol{k}}(\boldsymbol{r})$'s for $\{\Delta\boldsymbol{R}_{l,j} = 0\}$, which belongs to the same band index $n = \sigma$ (Fig. 1.7). That is, a set of $\{\psi_{\sigma,l,\{\Delta\boldsymbol{R}\}}(\boldsymbol{r})\}$ and a set of $\{\psi_{\sigma,k}(\boldsymbol{r})\}$ are related with each other by a unitary transformation. Then the **total** electronic ground state $\Psi^{N_e}_{g,\{\boldsymbol{R}\neq\bar{\boldsymbol{R}}\}}$ for $\{\Delta\boldsymbol{R}_{l,j} \neq 0\}$, i.e., $\{\boldsymbol{R} \neq \bar{\boldsymbol{R}}\}$,

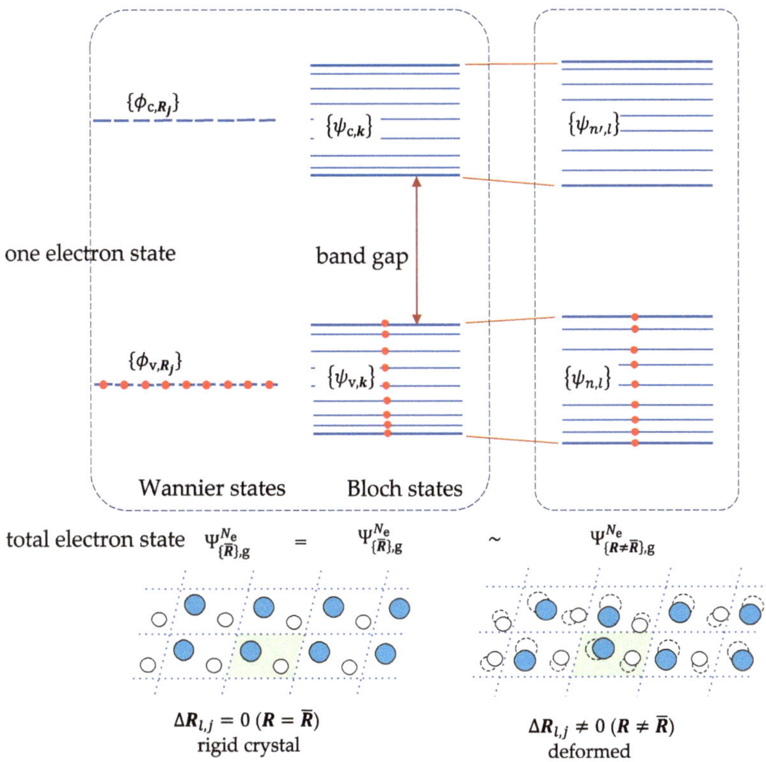

Figure 1.13 Modification of one-electron states in the electronic ground state of a semiconductor when $\{\Delta\boldsymbol{R}_{l,j} = 0\}$ and $\{\Delta\boldsymbol{R}_{l,j} \neq 0\}$.

$$\Psi^{N_e}_{\{\boldsymbol{R}\neq\bar{\boldsymbol{R}}\},g} = \frac{1}{\sqrt{N_e!}}\begin{vmatrix} \psi_{v,l_1,\{\Delta\boldsymbol{R}\}}(\boldsymbol{r}_1) & \psi_{v,l_1,\{\Delta\boldsymbol{R}\}}(\boldsymbol{r}_2) & \psi_{v,l_1,\{\Delta\boldsymbol{R}\}}(\boldsymbol{r}_3) & \cdots \\ \psi_{v,l_2,\{\Delta\boldsymbol{R}\}}(\boldsymbol{r}_1) & \psi_{v,l_2,\{\Delta\boldsymbol{R}\}}(\boldsymbol{r}_2) & \psi_{v,l_2,\{\Delta\boldsymbol{R}\}}(\boldsymbol{r}_3) & \cdots \\ \psi_{v,l_3,\{\Delta\boldsymbol{R}\}}(\boldsymbol{r}_1) & \psi_{v,l_3,\{\Delta\boldsymbol{R}\}}(\boldsymbol{r}_2) & \psi_{v,l_3,\{\Delta\boldsymbol{R}\}}(\boldsymbol{r}_3) & \cdots \\ \cdots & \cdots & \cdots & \cdots \end{vmatrix}, \tag{1.37a}$$

is equal to the total electronic ground state $\Psi^{N_e}_{g,\{\bar{R}\}}$ for $\{\Delta \boldsymbol{R}_{l,j} = 0\}$, $\{\boldsymbol{R} = \bar{\boldsymbol{R}}\}$.

$$\Psi^{N_e}_{\{\boldsymbol{R}\neq\bar{\boldsymbol{R}}\},g} = \frac{1}{\sqrt{N_e!}}\begin{vmatrix} \psi_{v,\boldsymbol{k}_1}(\boldsymbol{r}_1) & \psi_{v,\boldsymbol{k}_1}(\boldsymbol{r}_2) & \psi_{v,\boldsymbol{k}_1}(\boldsymbol{r}_3) & \cdots \\ \psi_{v,\boldsymbol{k}_2}(\boldsymbol{r}_1) & \psi_{v,\boldsymbol{k}_2}(\boldsymbol{r}_2) & \psi_{v,\boldsymbol{k}_2}(\boldsymbol{r}_3) & \cdots \\ \psi_{v,\boldsymbol{k}_3}(\boldsymbol{r}_1) & \psi_{v,\boldsymbol{k}_3}(\boldsymbol{r}_2) & \psi_{v,\boldsymbol{k}_3}(\boldsymbol{r}_3) & \cdots \\ \cdots & \cdots & \cdots & \cdots \end{vmatrix} = \Psi^{N_e}_{\{\bar{\boldsymbol{R}}\},g}, \tag{1.37b}$$

as far as all $\{\psi_{v,\boldsymbol{l}\{\Delta \boldsymbol{R}\}}(\boldsymbol{r})\}$ states are filled with N_e electrons. This is true for any pattern of small displacements $\{\Delta \boldsymbol{R}_{l,j}\}$. In other words, in the electronic ground state in semiconductors and insulators, occupied one-electron states may be changed and mixed together within the valence band subspace, but nothing changes in the total N_e-electron ground state. See also the equivalency of Eqs. (1.16) and (1.18).

It would be informative to remark that because the character of Wannier orbitals resembles that of atomic orbitals of the constituent ions, then we may consider approximately that the valence electrons follow the displacement of ions. So, hereafter we call "atom" instead of ion, and their cooperative motions the lattice vibrations.

1.5.6 Hamiltonian of the Lattice Vibrations

In order to discuss the dynamics of the lattice vibrations, it would be effective to introduce the Hamiltonian scheme. Using a new variable $q_{\sigma,\boldsymbol{k}} = f_{\sigma,\boldsymbol{k}} \exp[-i\omega_{\sigma,\boldsymbol{k}}t]$ including the time dependence in Eq. (1.36), the $\alpha(=x,y,z)$ component of the displacement of j-th ion in $\boldsymbol{l}$-th unit cell is given by

$$\Delta R^{\alpha}_{l,j} = \frac{1}{\sqrt{NM_j}}\sum_{\sigma,\boldsymbol{k}} e^{\alpha}_{\sigma,j,\boldsymbol{k}} q_{\sigma,\boldsymbol{k}} \exp[i\,\boldsymbol{k}\cdot\boldsymbol{T}_l], \quad \alpha = x,y,z. \tag{1.38}$$

Since $\Delta R^{\alpha}_{l,j}$ is a real quantity, $q^*_{\sigma,\boldsymbol{k}} = q_{\sigma,-\boldsymbol{k}}$. Put $\Delta R^{\alpha}_{l,j}$ of Eq. (1.38) into Eq. (1.27) with Eq. (1.28), after some cumbersome procedures, the Lagrangian has a form

$$L = \frac{1}{2}\sum_{\sigma,\boldsymbol{k}} (\dot{q}^*_{\sigma,\boldsymbol{k}}\dot{q}_{\sigma,\boldsymbol{k}} - \omega^2_{\sigma,\boldsymbol{k}} q^*_{\sigma,\boldsymbol{k}} q_{\sigma,\boldsymbol{k}}). \tag{1.39}$$

Next introducing the conjugate momentum

$$p_{\sigma,\boldsymbol{k}} = \frac{\partial L}{\partial \dot{q}_{\sigma,\boldsymbol{k}}} = \dot{q}^{*}_{\sigma,\boldsymbol{k}}, \tag{1.40}$$

the Hamiltonian (the total energy) of the system has a form

$$H_{\mathrm{L}} = \frac{1}{2}\sum_{\sigma,\boldsymbol{k}} (p^{*}_{\sigma,\boldsymbol{k}} p_{\sigma,\boldsymbol{k}} + \omega^{2}_{\sigma,\boldsymbol{k}} q^{*}_{\sigma,\boldsymbol{k}} q_{\sigma,\boldsymbol{k}}). \tag{1.41}$$

Now the collective atomic motions in a crystal is shown to be a set of independent harmonic oscillators, which are characterized by the branch index σ and the wave vector $\boldsymbol{k}$ with the angular frequency $\omega_{\sigma,\boldsymbol{k}}$, given by the solution of the secular Eq. (1.33). Thus, the set of $\{\omega_{\sigma,\boldsymbol{k}}\}$ stands for the vibrational spectrum of the material and reflects all the interactions V_{e}, V_{I}, and V_{eI} in it. In other words, in the ground state of a material, the interaction between many valence electrons and ions in scheme (B) results in the lattice vibration H_{L}. There is no "electron–lattice interaction" in the ground state. One may have a simple question: Since one-electron states $\psi_{\sigma,\boldsymbol{k}}(\boldsymbol{r})$ are defined at a given crystal structure, $\{\overline{\boldsymbol{R}}\}_{\mathrm{g}}$, then if some atoms are displaced from that, one-electron states would be affected and then the character will be changed, sometimes results in localization: This will be precisely discussed in Chapter 2.

1.5.7 Lattice Dynamics: Phonon [Quantum Treatment]

In order to discuss the quantum nature of the lattice vibrations, it is straightforward to focus the relation between $q_{\sigma,\boldsymbol{k}}$ and its conjugate momentum $p_{\sigma,\boldsymbol{q}}$ for $(\sigma, \boldsymbol{k})$ mode: They have a following commutation relation in quantum mechanics:

$$[q_{\sigma,\boldsymbol{k}}, p_{\sigma',\boldsymbol{k}'}] \equiv q_{\sigma,\boldsymbol{k}}\, p_{\sigma',\boldsymbol{k}'} - p_{\sigma',\boldsymbol{k}'}\, q_{\sigma,\boldsymbol{k}} = i\hbar\, \delta_{\sigma\sigma'}\, \delta_{\boldsymbol{k}\boldsymbol{k}'}. \tag{1.42}$$

Let us introduce the creation operator $b^{+}_{\sigma,\boldsymbol{k}}$ and its conjugate annihilation operator $b_{\sigma,\boldsymbol{k}}$,

$$b^{+}_{\sigma,\boldsymbol{k}} = \sqrt{\frac{\omega_{\sigma,\boldsymbol{k}}}{2\hbar}}\left(q_{\sigma,-\boldsymbol{k}} - \frac{i\, p_{\sigma,\boldsymbol{k}}}{\omega_{\sigma,\boldsymbol{k}}}\right), \quad b_{\sigma,\boldsymbol{k}} = \sqrt{\frac{\omega_{\sigma,\boldsymbol{k}}}{2\hbar}}\left(q_{\sigma,\boldsymbol{k}} + \frac{i\, p_{\sigma,-\boldsymbol{k}}}{\omega_{\sigma,\boldsymbol{k}}}\right). \tag{1.43a}$$

Or inversely,

$$q_{\sigma,k} = \sqrt{\frac{\hbar}{2\omega_{\sigma,k}}}(b_{\sigma,k}+b_{\sigma,-k}^{+}), \quad p_{\sigma,k} = i\sqrt{\frac{\hbar\omega_{\sigma,k}}{2}}(b_{\sigma,k}^{+} - b_{\sigma,-k}). \tag{1.43b}$$

According to Eq. (1.42), $b_{\sigma,k}$ and $b_{\sigma,k}^{+}$ obey the following commutation relations for Bose particles:

$$[b_{\sigma,k}, b_{\sigma',k'}^{+}] \equiv b_{\sigma,k}\, b_{\sigma',k'}^{+} - b_{\sigma',k'}^{+} b_{\sigma,k} = \delta_{\sigma\sigma'}\,\delta_{kk''},$$

$$[b_{\sigma,k}, b_{\sigma',k'}] \equiv b_{\sigma,k}\, b_{\sigma',k'} - b_{\sigma',k'}\, b_{\sigma,k} = 0,$$

$$[b_{\sigma,k}^{+}, b_{\sigma',k'}^{+}] \equiv b_{\sigma,k}^{+}\, b_{\sigma',k'}^{+} - b_{\sigma',k'}^{+} b_{\sigma,k}^{+} = 0. \tag{1.44}$$

In terms of $b_{\sigma,k}$ and $b_{\sigma,k}^{+}$, the Hamiltonian of the lattice are expressed as

$$H_{\rm L} = \sum_{\sigma,k} \hbar\omega_{\sigma,k}\left(b_{\sigma,k}^{+} b_{\sigma,k} + \frac{1}{2}\right). \tag{1.45}$$

The eigenstates and eigenenergy of $H_{\rm L}$ are given by

$$|n_{\sigma1,k1}, n_{\sigma2,k2}, \dots\rangle = \prod_{\sigma,k} \frac{1}{\sqrt{n_{\sigma,k}!}} (b_{\sigma,k}^{+})^{n_{\sigma,k}}\, |\mathbf{0}\rangle, \qquad |\mathbf{0}\rangle \equiv |0, 0, \dots\rangle \tag{1.46}$$

$$E = \sum_{\sigma,k} \hbar\omega_{\sigma,k}\left(n_{\sigma,k} + \frac{1}{2}\right). \qquad n_{\sigma,k} = 0, 1, 2, \dots \tag{1.47}$$

Here $|\mathbf{0}\rangle$ is the phonon vacuum state. The eigenenergies are equally separated by a quantum energy $\hbar\omega_{\sigma,k}$. There are following relations:

$$b_{\sigma,k}^{+}|\dots, n_{\sigma,k}, \dots\rangle = \sqrt{n_{\sigma,k} + 1}\,|\dots, n_{\sigma,k} + 1, \dots\rangle, \tag{1.48a}$$

$$b_{\sigma,k}|\dots, n_{\sigma,k}, \dots\rangle = \sqrt{n_{\sigma,k}}\,|\dots, n_{\sigma,k} - 1, \dots\rangle. \tag{1.48b}$$

Thus, we can call $b_{\sigma,k}^{+}$ the phonon creation operator and $b_{\sigma,k}$ the phonon annihilation operator for $(\sigma, \boldsymbol{k})$ mode. The state (1.46) describe a situation there are $n_{\sigma,\boldsymbol{k}}$ number phonons, each has an energy $\hbar\omega_{\sigma,\boldsymbol{k}}$.

The displacement $\Delta R_{l,j}^{\alpha}$ in quantum presentation is shown to have a form with phonon operators

$$\Delta R_{l,j}^{\alpha} = \sum_{\sigma,\boldsymbol{k}} e_{\sigma,j,\boldsymbol{k}}^{\alpha} \sqrt{\frac{\hbar}{2N\, M_j \omega_{\sigma,\boldsymbol{k}}}} \left(b_{\sigma,\boldsymbol{k}} + b_{\sigma,-\boldsymbol{k}}^{+}\right) \exp[i\, \boldsymbol{k} \cdot \boldsymbol{T}_l],$$

$$\alpha = x, y, z. \tag{1.49}$$

1.5.8 Thermal Properties of Phonons

When temperature is raised, a number of phonons are thermally excited. The average number of phonons at absolute temperature T [K] is given by the Bose–Einstein distribution

$$\langle n_{\sigma,\boldsymbol{k}} \rangle = \frac{1}{\exp\left[\hbar\omega_{\sigma,\boldsymbol{k}}/k_{\mathrm{B}}T\right] - 1}. \tag{1.50}$$

The thermal average of the energy of the total phonon (lattice vibration) systems is then given by

$$E(T) = \langle E \rangle_T = \sum_{\sigma,\boldsymbol{k}} \hbar\omega_{\sigma,\boldsymbol{k}} \left(\langle n_{\sigma,\boldsymbol{k}} \rangle + \frac{1}{2} \right). \tag{1.51}$$

With the aid of the phonon energy density of states,

$$D(\varepsilon) = \sum_{\sigma,\boldsymbol{k}} \delta\left(\varepsilon - \hbar\omega_{\sigma,\boldsymbol{k}}\right), \tag{1.52}$$

$$E(T) = \int_0^{\infty} D(\varepsilon) \frac{\varepsilon}{2} \frac{\exp[\varepsilon/kT] + 1}{\exp[\varepsilon/kT] - 1}\, d\varepsilon. \tag{1.53}$$

Thus, the phonon density of states $D(\varepsilon)$ is an important quantity that characterizes the thermal properties of condensed matter.

1.5.9 Specific Heat

The total energy $E(T)$ of the system is an important quantity; however, it is difficult to measure it directly. Only we can do is to measure an increase (or decrease) of the total energy by changing the circumstances, such as temperature, pressure, etc. Thus, the specific heat, c, of the system is an important parameter of the thermal properties of materials.

$$c = \frac{\partial}{\partial T} E(T) = \int_0^\infty D(\varepsilon) \frac{\partial}{\partial T} \left[\frac{1}{\exp[\varepsilon / k_B T] - 1} \right] d\varepsilon$$

$$= -\frac{1}{k_B T^2} \int_0^\infty D(\varepsilon) \frac{\exp\left[\frac{\varepsilon}{k_B T}\right]}{\left(\exp\left[\frac{\varepsilon}{k_B T}\right] - 1\right)^2} d\varepsilon. \quad (1.54)$$

The behavior of $D(\varepsilon)$ in lower energy is crucial for the temperature dependence of the specific heat. Once the phonon density of states $D(\varepsilon)$ is obtained, calculation of Eq. (1.54) is easily done. There are two typical models for a simple form of the density of states $D(\varepsilon)$.

(a) Einstein model

All phonon frequencies are the same ω_{op}: It is a typical modeling for the optical phonons, called the Einstein model.

$$D(\varepsilon) = 3N\delta(\varepsilon - \hbar\omega_{op}). \quad (1.55)$$

Here N is the number of unit cells. The specific heat is given by

$$c = -\frac{3N}{k_B T^2} \frac{\exp\left[\frac{\hbar\omega_{op}}{k_B T}\right]}{\left(\exp\left[\frac{\hbar\omega_{op}}{k_B T}\right] - 1\right)^2}. \quad (1.56)$$

A typical temperature dependence is illustrated in Fig. 1.14. We can readily see that

$$c = 3Nk_B \text{ at high temperature } (k_B T \gg \hbar\omega_{op}), \quad (1.56a)$$

then so-called the Dulong–Petit law holds. As temperature decreases toward $T \to 0$, then $c \to 0$ very sharply.

$$c = \frac{3N}{k_B T^2} \exp\left[-\hbar\omega_{op}/k_B T\right] \quad \text{at low temperature } \left(k_B T \ll \hbar\omega_{op}\right). \tag{1.56b}$$

Overall temperature dependence for $0 \leftrightarrow \infty$ agrees between the Einstein model and experiments to some extent (Fig. 1.14). There is a discrepancy at low temperature which can be improved if we take account of low energy excitations in the phonon modes.

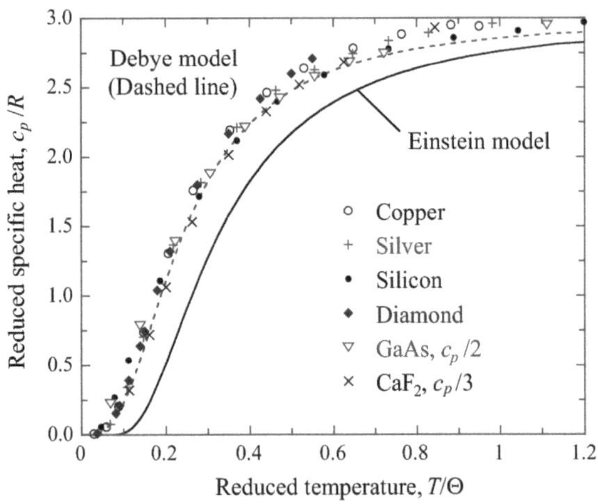

Figure 1.14 Temperature dependence of the specific heat in Einstein model, Debye model, and experimental data for several materials Reproduced with permission from Ref. [9].

(b) Debye model

This is a typical modeling for the acoustic phonons, called the Debye model. Assume a linear frequency dispersion,

$$\omega_{\sigma,\boldsymbol{k}} = u_\sigma k. \qquad \sigma = \text{L}, \text{T}. \tag{1.57}$$

Here, u_L and u_T are the sound velocity of the longitudinal and transverse modes, respectively. The phonon density of states is given by

$$D(\varepsilon) = \sum_{\sigma,\boldsymbol{k}} \delta\left(\varepsilon - \hbar\omega_{\sigma,\boldsymbol{k}}\right) = \begin{cases} \dfrac{V}{2\pi^2}\left(\dfrac{2}{u_T^3} + \dfrac{1}{u_L^3}\right)\omega^3 & \omega \le \omega_\text{D} \\ 0 & \omega > \omega_\text{D} \end{cases}. \tag{1.58}$$

Here, we introduce the cut-off frequency ω_D, wave number k_D and temperature T_D, as

$$\hbar\omega_\mathrm{D} \equiv \hbar u_\sigma k_\mathrm{D} \equiv k_\mathrm{B} T_\mathrm{D} \tag{1.59}$$

They are determined under the condition:

$$N = \frac{V}{(2\pi)^3}\iiint_{\text{1st Brillouin zone}} dk = \frac{V}{(2\pi)^3}\int_0^{k_\mathrm{D}} 4\pi k^2 dk = \int_0^{\hbar\omega_\mathrm{D}} D(\varepsilon)d\varepsilon.$$

Then $\omega_\mathrm{D} = \{18\pi^2 N/V(2u_T^{-3} + u_L^{-3})\}^{1/3}$. The specific heat is given by

$$c = \frac{9Nk_\mathrm{B}}{(\hbar\omega_\mathrm{D}/k_\mathrm{B}T)^3}\int_0^{\hbar\omega_\mathrm{D}/k_\mathrm{B}T} \frac{x^4}{(\mathrm{e}^x - 1)(1 - \mathrm{e}^{-x})} dx. \tag{1.60}$$

Materials dependence of the specific heat in the Debye model is factorized into a single parameter $\hbar\omega_\mathrm{D}/k_\mathrm{B}T$. A typical temperature dependence is illustrated in Fig. 1.14. We can readily see that at high temperature ($T \gg \hbar\omega_\mathrm{D}/k_\mathrm{B}$),

$$c = 3Nk_\mathrm{B}, \tag{1.60a}$$

then so-called the Dulong-Petit law holds.
At low temperature ($k_\mathrm{B}T \ll \hbar\omega_\mathrm{D}$),

$$c = 3Nk_\mathrm{B}\frac{4\pi^4}{5}\left(\frac{T}{T_\mathrm{D}}\right)^3. \tag{1.60b}$$

This T^3 dependence is known as the Debye law. Table 1.1 shows the Debye frequency ω_D and T_D in typical semiconductors [10]. The reason why the Debye model reasonably explains the temperature dependence of various materials is as follows: At low temperature, only low-frequency phonons are excited thermally, which have long wavelength. Most materials in large scale can be considered as an elastic continuum medium which does not reflect detailed structure of the unit cell.

Table 1.1 The Debye frequency ω_D and temperature T_D in low-temperature limit for several semiconductors, based on [10]

	C	Si	Ge	AlP	GaAs	ZnSe	CdTe
ω_D [10^{13} rad/s]	29.0	8.53	4.89	7.69	4.50	3.60	2.07
T_D [K]	2220	652	374	588	344	275	158

What happens when temperature is high as $\hbar\omega_D < k_B T \ll E_g$ (the electron band gap energy). All electrons still stay in the valence bands, but atomic motions can be thermally excited, and sometimes happen to overcome the potential barrier for reactions which have been neglected in the harmonic approximation Eq. (1.27). Thus the thermal melting of a crystal structure will take place when the average atomic displacement is roughly equal to the lattice constant, or the thermal energy $k_B T$ is equal to the potential barrier height ΔU for atomic displacements. This thermal melting is beyond the scope of this book. However, even at low temperature, nonthermal local melting could be realized by electronic excitations. This will be discussed in Chapter 9.

References

1. Shinozuka, Y., *Jpn. J. Appl. Phys.*, **32** (1993) 4560.
2. Kittel, C., *Introduction to Solid State Physics* 2nd ed. (Willey, 1956).
3. Ashcroft, N. W., and Mermin, N. D., *Solid State Physics* (Holt, Rinehart and Winston, 1976).
4. Yu, P. Y., and Cardona, M., *Fundamentals of Semiconductors* (Springer-Verlag, 1996).
5. Rössler, U., *Solid State Theory* (Springer-Verlag, 2004).
6. Ibach, H., and Luth, H., *Solid-State Physics: An Introduction to Principles of Materials Science*, (Springer, Berlin Heidelberg 2010).
7. *Springer Handbook of Condensed Matter and Materials Data* (Springer-Verlag, Berlin 2005).
8. Baroni, S., Stefano de Gironcoli, Dal Corso, A., and Giannozzi, P., *Rev. Mod. Phys.* **73** (2001) 515.
9. https://www.slideserve.com/base/thermal-properties-of-solids-and-the-size-effect.

10. Adachi, S., *Properties of Group-IV, III-V and II-VI Semiconductors* (John Wiley & Sons, Ltd, England, 2005), p. 30, Table 2.3.

Chapter 2

Electron–Lattice Interaction

When carriers (conduction electron, valence hole) are introduced to a semiconductor by a band-to-band photoexcitation or injection, the balance of the inter-atomic (ionic) forces mediated by valence electrons is disturbed and then the atomic configuration is rearranged. The induced atomic displacements in turn affect the electronic states. This is the origin of the electron–lattice interaction in nonmetallic materials [1].

2.1 Electronic Excited States

We have learned in the previous chapter that in a single-electron approximation, the electronic states $\psi_{\sigma,\boldsymbol{k}}(\boldsymbol{r})$ in a crystal where each ion is fixed at the regular position $\{\bar{\boldsymbol{R}}\}$ are characterized by the band index σ and the wave vector $\boldsymbol{k}$ (see Eq. (1.13)). In the ground state $\Psi^{N_e}_{g,\{\bar{R}\}}$, N_e electrons, respectively, occupy one-electron state $\psi_{v,\boldsymbol{k}}(\boldsymbol{r})$ with $\varepsilon_v(\boldsymbol{k})$ of the valence band. The energy of the ground state can be formally given by

$$E_{\text{gound}} = \left\langle \Psi^{N_e}_{g,\{\bar{\boldsymbol{R}}\}} \middle| H_e + V_{eI} + V_I \middle| \Psi^{N_e}_{g,\{\bar{\boldsymbol{R}}\}} \right\rangle. \tag{2.1}$$

Here the kinetic energy of ions is omitted. The value of E_{ground} can be evaluated in principle, but unknown at present. So, let the origin

Electron–Lattice Interactions in Semiconductors
Yuzo Shinozuka

ISBN 978-981-4800-96-9 (Hardcover), 978-1-003-02346-3 (eBook)
www.jennystanford.com

of energy be $E_{\text{ground}} = 0$. In semiconductors and insulators, there is an energy gap between the valence and conduction bands. An excitation of an electron from $\psi_{\text{v},\boldsymbol{k}_\text{h}}(\boldsymbol{r})$ in the valence band to $\psi_{\text{c},\boldsymbol{k}_\text{e}}(\boldsymbol{r})$ in the conduction band (Fig. 2.1a) needs an energy

$$E_{\text{v},\boldsymbol{k}_\text{h},\text{c},\boldsymbol{k}_\text{e}} = \varepsilon_{\text{c},\boldsymbol{k}_\text{e}} - \varepsilon_{\text{v},\boldsymbol{k}_\text{h}}. \tag{2.2}$$

This excited state has double continuum spectrum with respect to $\boldsymbol{k}_\text{h}$ in the valence band and $\boldsymbol{k}_\text{e}$ in the conduction band (Fig. 2.1b). One can regard this excited state as a pair of carriers: a conduction electron with energy $\varepsilon_{\text{c},\boldsymbol{k}_\text{e}}$ and a valence hole with energy $-\varepsilon_{\text{v},\boldsymbol{k}_\text{h}}$. The allocation of the energy to two carriers is arbitral, depending on the choice of the origin of one-electron energy $\varepsilon_\sigma(\boldsymbol{k})$, but does not affect the excitation energy (2.2).

If we apply the effective mass approximation for the band energies of carriers (details will be given in Section 2.2.2),

$$E_{\text{v},\boldsymbol{k}_\text{h},\text{c},\boldsymbol{k}_\text{e}} = E_\text{g} + \frac{\hbar^2\boldsymbol{k}_\text{e}^2}{2m_\text{e}^*} - \frac{\hbar^2\boldsymbol{k}_\text{h}^2}{2m_\text{h}^*}. \tag{2.3}$$

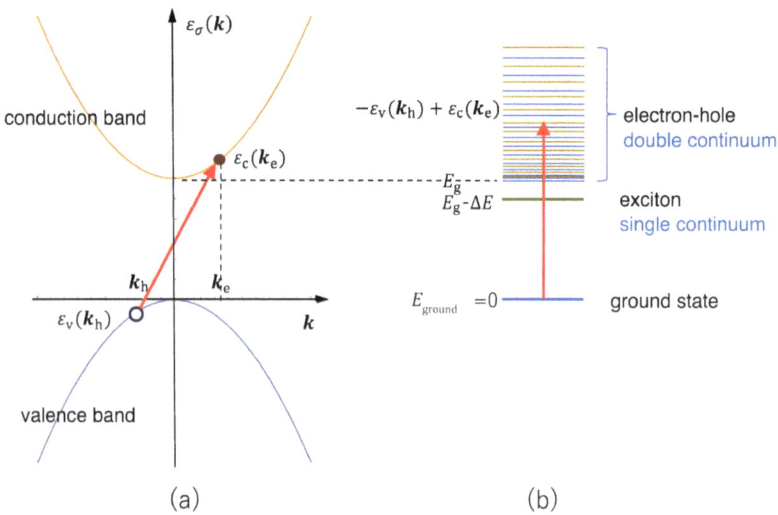

Figure 2.1 (a) One-electron states in a semiconductor as a function of a k-vector. (b) Energy diagram for excited states showing double continuum spectrum of an electron and a hole. One can put the exciton level into (b) but cannot put into (a)!

Here E_g is the (fundamental) band gap energy, and m_h^* and m_e^* are the effective mass of a carrier (hole and electron). If we further take account of the Coulomb attraction between a hole and an electron, there arises an exciton state just ΔE below E_g. The distance ΔE is called the exciton binding energy (Fig. 2.1b). It should be noted that one cannot put the exciton level in the one-electron scheme such as Fig. 2.1a: Otherwise it might lead us incorrect conclusion. In discussing exciton states, we should use a scheme similar to Fig. 2.1b.

In the following, we allow ions to move ($\boldsymbol{R}_j = \bar{\boldsymbol{R}}_j + \Delta\boldsymbol{R}_j$) and study the effects to electronic excited states: an electron in the conduction band and a hole in the valence band, first separately.

2.2 Hamiltonian of a Single Carrier

2.2.1 A Conduction Electron

Let us first consider a situation where an extra electron is added to a perfect crystal. For careful consideration, we will not use scheme (A) in Chapter 1, but use scheme (B) taking care of (N_e + 1) electrons. In principle, the situation can be investigated by calculating the adiabatic potentials $U_l^{N_e+1}(\{\Delta\boldsymbol{R}\})$ for the l-th state of (N_e + 1) electrons as a function of the atomic (ionic) displacements $\{\Delta\boldsymbol{R}\}$, which could be evaluated with a suitable first-principles calculation. As previously stated in Section 1.5, one-electron states $\{\psi_{\sigma,l\{\Delta\boldsymbol{R}\}}(\boldsymbol{r})\}$ and their energy $\{\varepsilon_{\sigma,l}(\{\Delta\boldsymbol{R}\})\}$ are affected by the ionic displacements $\{\Delta\boldsymbol{R}\}$. We have seen in the previous section, that as long as the displacement $\{\Delta\boldsymbol{R}\}$ are small, a set of electronic state $\psi_{\sigma,l\{\Delta\boldsymbol{R}\}}(\boldsymbol{r})$'s in a deformed configuration $\{\Delta\boldsymbol{R} \neq 0\}$ are constructed from a linear combination of Bloch state $\psi_{\sigma,\boldsymbol{k}}(\boldsymbol{r})$'s in $\{\Delta\boldsymbol{R} = 0\}$, both belong to the same band index σ. In order to make physics clear, let us focus on the following situation (Fig. 2.2):

(a) Electronic state:

N_e electrons occupy the lowest possible one-electron states originated from the valence band. As can be seen from Fig. 2.2, as long as these states are completely filled, the quantum number "p" for the (N_e + 1) electron system is assigned to the freedom of an extra electron. An extra electron occupies the p-th one-electron state $\psi_{c,p\{\Delta\boldsymbol{R}\}}(\boldsymbol{r})$ originated from the conduction band with an energy $\varepsilon_{c,p}(\{\Delta\boldsymbol{R}\})$ which depends on the atomic displacements $\{\Delta\boldsymbol{R}\}$.

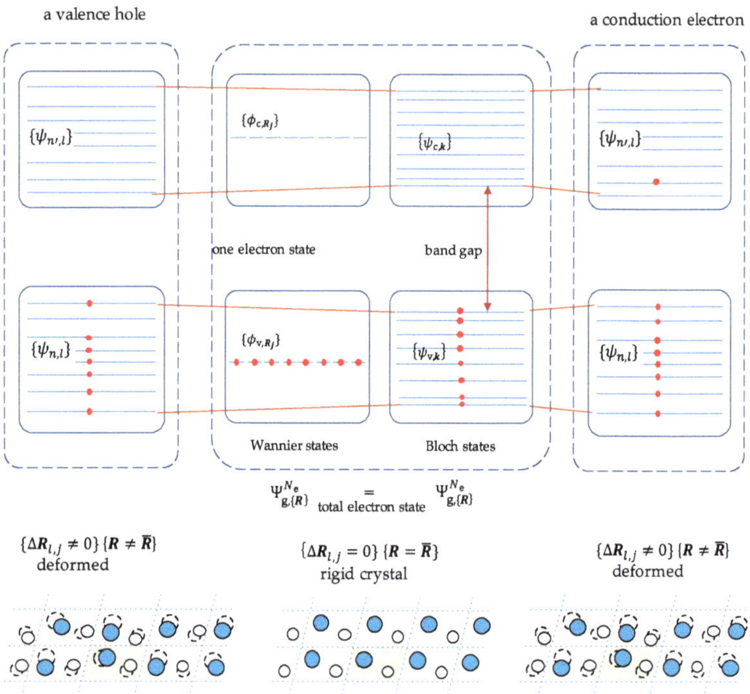

Figure 2.2 Several occupation patterns of the one-electron states for the rigid crystal $\{\Delta\boldsymbol{R} = 0\}$ and deformed crystal $\{\Delta\boldsymbol{R} \neq 0\}$.

To avoid unnecessary complexity in many-particle presentation, such as using the Slater determinant (1.16), we will use the second quantized form introduced in Section 1.4. Remember that the ground state $\Psi^{N_e}_{g,\{\bar{\boldsymbol{R}}\}}$ of N_e electrons in a crystal material can be expressed as a product of many creation operators operating to the (true) vacuum $|0\rangle$:

$$\left|\Psi^{N_e}_{g,\{\bar{\boldsymbol{R}}\}}\right\rangle = \prod_{\boldsymbol{k}=\text{occupied}} a^+_{v,\boldsymbol{k}}\,|0\rangle = \prod_{\boldsymbol{T}_l} a^+_{v,\boldsymbol{T}_l}\,|0\rangle \equiv |G0\rangle. \tag{2.4}$$

Here $a^+_{v,\boldsymbol{k}}$ and $a^+_{v,\boldsymbol{T}_l}$ are the creation operator for a Bloch orbital $\psi_{v,\boldsymbol{k}}(\boldsymbol{r})$ with a wave vector $\boldsymbol{k}$ and for a Wannier orbital

$\phi_{\mathrm{v},\boldsymbol{T}_l}(\boldsymbol{r})$ at unit cell $\boldsymbol{T}_l$, respectively. Both operators are related with a Fourier transformation:

$$a^{+}_{\mathrm{v},\boldsymbol{T}_l} = \frac{1}{\sqrt{N}}\sum_{\boldsymbol{k}} \exp(-i\boldsymbol{k}\cdot\boldsymbol{T}_l)\, a^{+}_{\mathrm{v},\boldsymbol{k}}. \tag{2.5a}$$

$$a_{\mathrm{v},\boldsymbol{T}_l} = \frac{1}{\sqrt{N}}\sum_{\boldsymbol{k}} \exp(-i\boldsymbol{k}\cdot\boldsymbol{T}_l) a_{\mathrm{v},\boldsymbol{k}}. \tag{2.5b}$$

It would be effective to rename this ground state as "vacuum", $|\mathrm{G0}\rangle$, that represents the ground state of N_e electron system with no "carrier."

One-electron state $\psi_{\mathrm{c},p\{\Delta\boldsymbol{R}\}}(\boldsymbol{r})$, whose details are unknown at present, can be expressed either by a linear combination of the Bloch state $\{\psi_{\mathrm{c},\boldsymbol{k}_\mathrm{e}}(\boldsymbol{r})\}$ or Wannier state $\{\phi_{\mathrm{c},\boldsymbol{T}_l}(\boldsymbol{r})\}$ in the conduction band. Thus, the present $N_\mathrm{e} + 1$ electron state can be expressed as a linear combination of either

$$a^{+}_{\mathrm{c},\boldsymbol{k}_\mathrm{e}}|\mathrm{G0}\rangle \;\leftrightarrow\; \text{an electron in } \psi_{\mathrm{c},\boldsymbol{k}_\mathrm{e}} + \text{fully occupied valence band.} \tag{2.6a}$$

or

$$a^{+}_{\mathrm{c},\boldsymbol{T}_l}|\mathrm{G0}\rangle \;\leftrightarrow\; \text{an electron in } \phi_{\mathrm{c},\boldsymbol{T}_l} + \text{fully occupied valence band.} \tag{2.6b}$$

(b) Adiabatic potentials:

In the Hartree–Fock approximation, Koopmans' theorem states that

$$U_l^{N_\mathrm{e}+1}(\{\Delta\boldsymbol{R}\}) = \varepsilon_p(\{\Delta\boldsymbol{R}\}) + U_\mathrm{g}^{N_\mathrm{e}}(\{\Delta\boldsymbol{R}\}). \tag{2.7}$$

Here, $U_\mathrm{g}^{N_\mathrm{e}}(\{\Delta\boldsymbol{R}\})$ is the adiabatic potential for the ground state of N_e electrons, which reflects all the interaction among ions and valence electrons, resulting in the so-called lattice elastic energy. As seen in Chapter 1, $U_\mathrm{g}^{N_\mathrm{e}}(\{\Delta\boldsymbol{R}\})$ can be diagonalized with use of the normal modes $\{\omega_{\sigma,\boldsymbol{k}}\}$, $\{q_{\sigma,\boldsymbol{k}}\}$ of the lattice vibrations in the harmonic approximation: σ denotes the mode and $\boldsymbol{k}$ stands for the wave vector,

$$U_\mathrm{g}^{N_\mathrm{e}}(\{\Delta\boldsymbol{R}\}) = \text{const.}\,(=0) + \frac{1}{2}\sum_{\sigma,\boldsymbol{k}} \omega^2_{\sigma,\boldsymbol{k}} q^*_{\sigma,\boldsymbol{k}} q_{\sigma,\boldsymbol{k}}. \tag{2.8}$$

Adding the kinetic energy of ions, we obtain the Hamiltonian of the lattice vibrations (phonons): See Section 1.5.

$$H_{\mathrm{L}} = \frac{1}{2}\sum_{\sigma,\boldsymbol{k}} (p^*_{\sigma,\boldsymbol{k}} p_{\sigma,\boldsymbol{k}} + \omega^2_{\sigma,\boldsymbol{k}} q^*_{\sigma,\boldsymbol{k}} q_{\sigma,\boldsymbol{k}}). \tag{2.9}$$

Thus, we can safely focus on one-electron $\psi_{\mathrm{c},p\{\Delta R\}}(\boldsymbol{r})$ or $\psi_{\mathrm{c},p\{q_k\}}(\boldsymbol{r})$ in the conduction band in addition to the elastic lattice.

2.2.2 Effective Hamiltonian for Conduction Electron

In a perfect periodic crystal with $\{\Delta \boldsymbol{R} = 0\}$, a conduction electron feels the periodic potential, $V_{\mathrm{eff}}(\boldsymbol{r}) = V_{\mathrm{eff}}(\boldsymbol{r} + \boldsymbol{T}_l)$ (Eq. (1.12)), originated from the interaction with ions and other electrons.

$$h^{\mathrm{eff}}_{\mathrm{e}}\, \psi_{\mathrm{c},\boldsymbol{k}}(\boldsymbol{r}) = \varepsilon_{\mathrm{c}}(\boldsymbol{k}) \psi_{\mathrm{c},\boldsymbol{k}}(\boldsymbol{r}): \quad h^{\mathrm{eff}}_{\mathrm{e}} \equiv \frac{\boldsymbol{p}^2}{2m} + V_{\mathrm{eff}}(\boldsymbol{r}), \tag{2.10}$$

where m is the true electron mass. The eigenstates are the Bloch states $\psi_{\mathrm{c},k}(\boldsymbol{r})$ (see Section 1.4), which can be expressed by a linear combination of Wannier state $\phi_{\mathrm{c},\boldsymbol{T}_l}$ localized at $\boldsymbol{T}_l$ (Eq. (1.17)).

$$\psi_{\mathrm{c},\boldsymbol{k}}(\boldsymbol{r}) = \frac{1}{\sqrt{N}} \sum_{\boldsymbol{T}_l} \exp(i\boldsymbol{k} \cdot \boldsymbol{T}_l) \phi_{\mathrm{c},\boldsymbol{T}_l}(\boldsymbol{r}), \tag{2.11a}$$

$$\phi_{\mathrm{c},\boldsymbol{T}_l}(\boldsymbol{r}) = \frac{1}{\sqrt{N}} \sum_{\boldsymbol{k}} \exp(-i\boldsymbol{k} \cdot \boldsymbol{T}_l)\, \psi_{\mathrm{c},\boldsymbol{k}}(\boldsymbol{r}). \tag{2.11b}$$

Let add a potential $h'(\boldsymbol{r})$ that varies gradually in space, originated from the lattice displacements $\{\Delta \boldsymbol{R} \neq 0\}$ and/or crystal imperfections (impurity, vacancy, ...).

$$\frac{\boldsymbol{p}^2}{2m} + V_{\mathrm{eff}}(\boldsymbol{r}) + h'(\boldsymbol{r}) \equiv h^{\mathrm{eff}}_{\mathrm{e}} + h'. \tag{2.12}$$

Our task is to obtain a tractable form of the Schrödinger equation for $h^{\mathrm{eff}}_{\mathrm{e}} + h'$.

$$(h^{\mathrm{eff}}_{\mathrm{e}} + h')\, \Psi_{\mathrm{c}}(\boldsymbol{r}) = E\, \Psi_{\mathrm{c}}(\boldsymbol{r}). \tag{2.13}$$

The eigenstate $\Psi_c(\boldsymbol{r})$ can be expressed with a set of Wannier states $\phi_{c,\boldsymbol{T}_l}$ as,

$$\Psi_c(\boldsymbol{r}) = \frac{1}{\sqrt{N}}\sum_{\boldsymbol{T}_l} F(\boldsymbol{T}_l)\phi_{c,\boldsymbol{T}_l}(\boldsymbol{r}). \tag{2.14}$$

Let us find the unknown coefficient $F(\boldsymbol{T}_l)$, i.e., find an equation for $F(\boldsymbol{T}_l)$.

The matrix elements of h_e^{eff} and h' in the Wannier base are

$$\langle h_e^{\text{eff}}\rangle_{ll'} \equiv \left\langle \phi_{c,\boldsymbol{T}_l}\middle| h_e^{\text{eff}} \middle| \phi_{c,\boldsymbol{T}_{l'}}\right\rangle = \frac{1}{N}\sum_{\boldsymbol{k}} \exp[-i\boldsymbol{k}\cdot(\boldsymbol{T}_l - \boldsymbol{T}_{l'})]\,\varepsilon_c(\boldsymbol{k}), \tag{2.15a}$$

$$\langle h'\rangle_{ll'} \equiv \left\langle \phi_{c,\boldsymbol{T}_l}\middle| h' \middle| \phi_{c,\boldsymbol{T}_{l'}}\right\rangle = \int \phi^*_{c,\boldsymbol{T}_l}(\boldsymbol{r})h'(\boldsymbol{r})\phi_{c,\boldsymbol{T}_l}(\boldsymbol{r})d\boldsymbol{r} = \langle h'\rangle_{ll}\delta_{ll'}. \tag{2.15b}$$

We have used that $h'(\boldsymbol{r})$ varies gradually in space and two Wannier orbitals at different site are orthogonal. Put Eq. (2.14) into (2.13), the matrix form of Eq. (2.13) has turned out to be

$$\sum_{\boldsymbol{T}_{l'}}\left(\langle h_e^{\text{eff}}\rangle_{ll'} + \langle h'\rangle_{ll'}\delta_{ll'} - E\delta_{ll'}\right)F(\boldsymbol{T}_{l'}) = 0. \tag{2.16}$$

The first term can be rewritten as

$$\sum_{\boldsymbol{T}_{l'}}\langle h_e^{\text{eff}}\rangle_{ll'}F(\boldsymbol{T}_{l'}) = \sum_{\boldsymbol{T}_m}\langle h_e^{\text{eff}}\rangle_{m0}F(\boldsymbol{T}_l - \boldsymbol{T}_m).$$

Using the Taylor expansion

$$\begin{aligned}F(\boldsymbol{T}_l - \boldsymbol{T}_m) &= F(\boldsymbol{T}_l) + (-\boldsymbol{T}_m)\frac{d}{dT}F(T) + \frac{1}{2!}(-\boldsymbol{T}_m)^2\frac{d^2}{dT^2}F(T) + \cdots \\ &= \exp\left[-\boldsymbol{T}_m\frac{d}{dT}\right]F(\boldsymbol{T}_l),\end{aligned}$$

we find

$$\sum_{\boldsymbol{T}_{l'}}\langle h_e^{\text{eff}}\rangle_{ll'}F(\boldsymbol{T}_{l'}) = \sum_{\boldsymbol{T}_m}\langle h_e^{\text{eff}}\rangle_{m0}F(\boldsymbol{T}_l - \boldsymbol{T}_m). \tag{2.17}$$

Here, we have used the inverse Fourier transform of Eq. (2.15a) with replacing $\boldsymbol{k}$ by $-i\nabla$. Next, let expand the energy $\varepsilon_c(\boldsymbol{k})$ around the bottom at $\boldsymbol{k}_{\text{min}}$,

$$\varepsilon_c(\boldsymbol{k}) = \varepsilon_c(\boldsymbol{k}_{\text{min}}) + \frac{\hbar^2(\boldsymbol{k}-\boldsymbol{k}_{\text{min}})^2}{2m_e^*}. \tag{2.18}$$

Here m_e^* is the effective mass of an electron, defined by

$$\frac{1}{m_e^*} = \frac{1}{\hbar^2}\frac{\partial^2\varepsilon_c(\boldsymbol{k})}{\partial k^2}\bigg|_{\text{min}}, \tag{2.19}$$

If crystals are not isotropic, the effective mass has a tensor quantity. The function $F(\boldsymbol{T}_l)$ is originally defined only at discrete lattice points $\boldsymbol{T}_l$, but it can be interpolated to be an envelope function $F(\boldsymbol{R})$ of a continuous variable $\boldsymbol{R}$ (Fig. 2.3). As for the normalization,

$$\sum_{\boldsymbol{T}_l}|F(\boldsymbol{T}_l)|^2 = 1 \ \ (\text{discrete } \boldsymbol{T}_l), \quad \int|F(\boldsymbol{R})|^2 d\boldsymbol{R} = 1 \ \ (\text{continuous } \boldsymbol{R}). \tag{2.20}$$

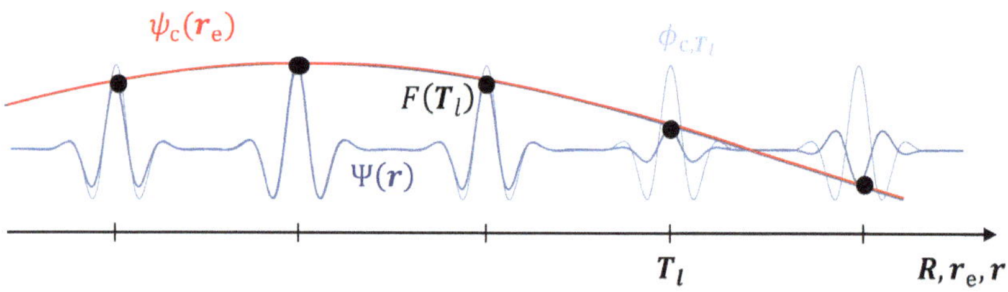

Figure 2.3 Electron wave function is described as an Envelope function $\psi_c(\boldsymbol{r}_e)$ of Wannier orbitals $\phi_{c,\boldsymbol{T}_l}$. Note that $\psi_c(\boldsymbol{r}_e)$ is originally defined only at discrete points $\{\boldsymbol{T}_l\}$ as $F(\boldsymbol{T}_l)$.

Putting Eq. (2.17) into (2.16), we obtain an equation for the envelope function $F(\boldsymbol{R})$.

$$\left[-\frac{\hbar^2}{2m_e^*}\frac{\partial^2}{\partial \boldsymbol{R}^2} + h'(\boldsymbol{R})\right]F(\boldsymbol{R}) = E\,F(\boldsymbol{R}). \tag{2.21a}$$

The variable $\boldsymbol{R}$ is further replaced by $\boldsymbol{r}$, the position operator of a conduction electron, and $F(\boldsymbol{R}) \rightarrow \psi(\boldsymbol{r})$ is considered as a wave function of an electron, we finally have

$$\left[-\frac{\hbar^2}{2m_\mathrm{e}^*}\frac{\partial^2}{\partial \boldsymbol{r}^2} + h'(\boldsymbol{r})\right]\psi(\boldsymbol{r}) = E\,\psi(\boldsymbol{r}). \tag{2.21b}$$

Thus, we have obtained the Shrödinger equation in the effective mass approximation, i.e., $h_\mathrm{e}^\mathrm{eff}$ in Eq. (2.10) can be replaced by

$$h_\mathrm{e}^\mathrm{eff} \equiv \frac{\boldsymbol{p}^2}{2m} + V_\mathrm{eff}(\boldsymbol{r}) = -\frac{\hbar^2}{2m_\mathrm{e}^*}\frac{\partial^2}{\partial \boldsymbol{r}^2} = \frac{\boldsymbol{p}^2}{2m_\mathrm{e}^*}. \tag{2.22}$$

Finally, the second quantization form of Hamiltonian in the effective mass approximation of a carrier in σ = c, v band is given as

$$\hat{h}_\mathrm{e}^\mathrm{eff} = \sum_{\boldsymbol{k}} \varepsilon_\sigma(\boldsymbol{k})\, a_{\sigma,\boldsymbol{k}}^+ a_{\sigma,\boldsymbol{k}}, \quad \hat{h}' = \sum_{\boldsymbol{k},\boldsymbol{k}'} h'_{\boldsymbol{k}\boldsymbol{k}'}\, a_{\sigma,\boldsymbol{k}}^+ a_{\sigma,\boldsymbol{k}'} \tag{2.23}$$

here,

$$\langle \psi_{\sigma,\boldsymbol{k}} | \hat{h}_0 | \psi_{\sigma,\boldsymbol{k}'} \rangle = \varepsilon_\sigma(\boldsymbol{k})\delta_{\boldsymbol{k}\boldsymbol{k}'}, \quad \langle \psi_{\sigma,\boldsymbol{k}} | \hat{h}' | \psi_{\sigma,\boldsymbol{k}'} \rangle = h'_{\boldsymbol{k}\boldsymbol{k}'} \tag{2.23a}$$

$$\langle \psi_{\sigma,\boldsymbol{k}} | \psi_{\sigma,\boldsymbol{k}'} \rangle = \delta_{\boldsymbol{k}\boldsymbol{k}'}, \quad \langle \psi_{\sigma,\boldsymbol{k}} | \exp[i\boldsymbol{q}\cdot\boldsymbol{r}]\, | \psi_{\sigma,\boldsymbol{k}'} \rangle = \delta_{\boldsymbol{k}\,\boldsymbol{k}'+\boldsymbol{q}} \tag{2.23b}$$

$$\psi_{\sigma,\boldsymbol{k}}(\boldsymbol{r}) = \frac{1}{\sqrt{V}}\mathrm{e}^{i\boldsymbol{k}\cdot\boldsymbol{r}}, \qquad h'_{\boldsymbol{k}\boldsymbol{k}'} = \frac{1}{V}\int \mathrm{e}^{-i\boldsymbol{k}\cdot\boldsymbol{r}}\hat{h}'\,\mathrm{e}^{i\boldsymbol{k}'\cdot\boldsymbol{r}}\,d\boldsymbol{r}. \tag{2.23c}$$

2.2.3 Effective Hamiltonian for Interaction Part: V_eL

Now is the time to derive the form of $h'(\boldsymbol{r})$ originated from the interaction V_eL between an extra conduction electron and the lattice (=other valence electrons and ions). This can be done by the following procedure [2, 3]. Let us first recall the displacement of the j-th ion in $\boldsymbol{T}_l$ unit cell in Chapter 1 (Eqs. (1.38) and (1.49)),

$$\Delta R_{l,j}^{\alpha} = \frac{1}{\sqrt{NM_j}} \sum_{\sigma,k} e_{\sigma,j,k}^{\alpha} q_{\sigma,k} \exp[i\,\boldsymbol{k} \cdot \boldsymbol{T}_l], \quad \alpha = x, y, z$$

$$= \sum_{\sigma,k} e_{\sigma,j,k}^{\alpha} \sqrt{\frac{\hbar}{2N\,M_j \omega_{\sigma,k}}} (b_{\sigma,k} + b_{\sigma,-k}^{+}) \exp[i\,\boldsymbol{k} \cdot \boldsymbol{T}_l]$$

Suppose all atoms in the crystal displace along a normal mode $q_{\sigma,k}$ (Fig. 2.4), a conduction electron in a Wannier orbital at $\boldsymbol{T}_l$ feels an induced effective potential proportional to the lattice shift, $q_{\sigma,k} \exp(i\boldsymbol{k} \cdot \boldsymbol{T}_l)$. This is because the induced potential must have the same periodicity characterized by the phonon wave vector $\boldsymbol{k}$. Replacing the discrete variable $\boldsymbol{T}_l$ by a continuous variable $\boldsymbol{r}_e$ and combining the contributions from all the modes, we obtain a formula,

$$V_{eL}(\boldsymbol{r}_e) = \sum_{\sigma,k} A_{\sigma,k}^{c}\, q_{\sigma,k} \exp(i\,\boldsymbol{k} \cdot \boldsymbol{r}_e). \tag{2.24}$$

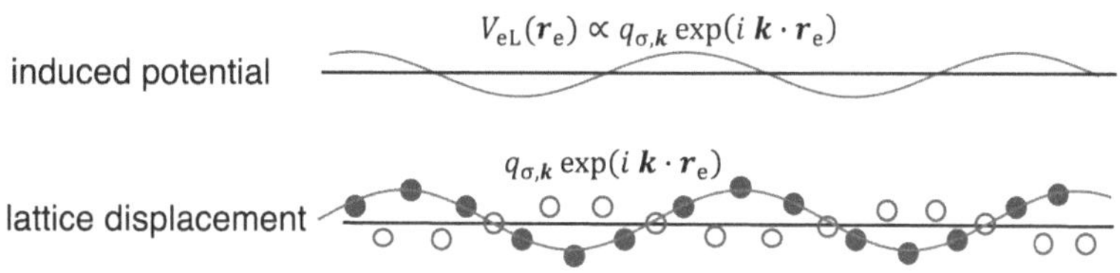

Figure 2.4 A normal lattice mode $q_{\sigma,k}$ with a wave vector $\boldsymbol{k}$ generates an induced potential $V_{eL}(\boldsymbol{r}_e)$ for a conduction electron at $\boldsymbol{r}_e = \boldsymbol{T}_l$ proportional to $q_{\sigma,k}\exp(i\boldsymbol{k} \cdot \boldsymbol{r}_e)$ with a phase difference.

The coefficient $A_{\sigma,k}^{c}$ depends on the phonon mode $(\sigma,\boldsymbol{k})$ and reflects the nature of the interaction, especially its force range. In the following, we derive the explicit form of $A_{\sigma,k}^{c}$ for two typical phonon modes in the long wave limit, where an electron can be treated by the effective mass approximation and the lattice vibrations by a continuum elastic medium.

Consider a generalized lattice distortion $D_\sigma(\boldsymbol{R})$ at $\boldsymbol{R}(=\boldsymbol{T}_l)$ created by the σ mode. It would be an ion displacement $\Delta R_{l,j}^{\alpha}$ (Eq. (1.38)) itself or the spatial derivative, ..., including the information within a unit cell. Assuming a scalar quantity at present, the Fourier component is

$$d_{\sigma,\boldsymbol{k}} = \frac{1}{\sqrt{N}} \sum_{\boldsymbol{R}} D_{\sigma}(\boldsymbol{R}) \exp[-i\, \boldsymbol{k} \cdot \boldsymbol{R}]. \tag{2.25}$$

The interaction $V_{\text{int}}(\boldsymbol{r}_{\text{e}} - \boldsymbol{R})$ between $D_{\sigma}(\boldsymbol{R})$ and a conduction electron at $\boldsymbol{r}_{\text{e}}$ (occupying a Wannier state $\phi_{c,\boldsymbol{T}_{l'}}$ at $\boldsymbol{T}_{l'} \rightarrow \boldsymbol{r}_{\text{e}}$) is a function of $\boldsymbol{r}_{\text{e}} - \boldsymbol{R}$, and is called the interaction kernel (Fig. 2.5). Summing up distortions at every lattice point $\boldsymbol{R} = \boldsymbol{T}_l$, an electron feels the potential

$$V_{\text{eL}}(\boldsymbol{r}_{\text{e}}) = \sum_{\boldsymbol{R}} V_{\text{int}}(\boldsymbol{r}_{\text{e}} - \boldsymbol{R})\, D_{\sigma}(\boldsymbol{R})$$

$$= \sum_{\boldsymbol{k}} v_{\text{int}}(\boldsymbol{k})\, d_{\sigma,\boldsymbol{k}} \exp[i\, \boldsymbol{k} \cdot \boldsymbol{r}_{\text{e}}], \tag{2.26}$$

where $v_{\text{int}}(\boldsymbol{k})$ is the Fourier component of $V_{\text{int}}(\boldsymbol{r})$:

$$V_{\text{int}}(\boldsymbol{r}) = \frac{1}{\sqrt{N}} \sum_{\boldsymbol{k}} v_{\text{int}}(\boldsymbol{k}) \exp[i\, \boldsymbol{k} \cdot \boldsymbol{r}]. \tag{2.27}$$

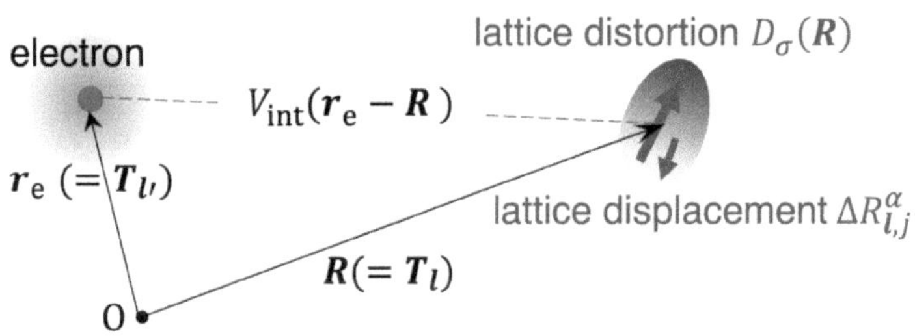

Figure 2.5 The lattice distortion $D_{\sigma}(\boldsymbol{R})$ at $\boldsymbol{R}$ (=$\boldsymbol{T}_l$) generates an additional potential $V_{\text{int}}(\boldsymbol{r}_{\text{e}} - \boldsymbol{R})$ to a conduction electron in a Wannier orbital at $\boldsymbol{T}_{l'}$(= $\boldsymbol{r}_{\text{e}}$).

Typical interactions induced by the LA mode and LO mode are

lattice mode σ	(1) LA	(2) LO
lattice distortion $D_{\sigma}(\boldsymbol{R})$	dilation	dipole moment
induced potential at $\boldsymbol{r}_{\text{e}}$	deformation potential	electronic dipole field

(1) Longitudinal acoustic (LA) mode

A longitudinal mode of the acoustic phonons generates a local lattice dilation in the long wave limit. The lattice dilation corresponds to a local increase or decrease of the volume of a unit cell (Fig. 2.6a). In the long wavelength limit, we can divide the crystal by a series of sub crystals whose lattice constant are different. Since the electronic band energy depends on the local lattice constant, difference in the lattice constant affects the electronic energy bands, resulting in a local energy shift of the bottom (top) of the conduction (valence) band, proportional to the dilation,

$$D_{\mathrm{LA}}(\boldsymbol{R}) = \mathrm{div}\Delta\boldsymbol{R}_l = i\boldsymbol{k}\cdot\Delta\boldsymbol{R}, \qquad \text{hence } d_{\mathrm{LA},\boldsymbol{k}} = \frac{i}{\sqrt{M}} k\, q_{\mathrm{LA},\boldsymbol{k}} \tag{2.28}$$

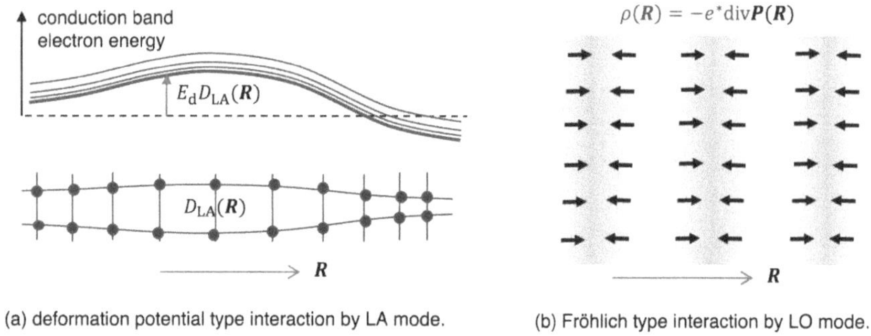

Figure 2.6 (a) A longitudinal acoustic phonon generates a local lattice dilation $\Delta(\boldsymbol{R})$, which makes a shift $E_{\mathrm{d}}^{\mathrm{c}}\Delta(\boldsymbol{R})$ in the conduction band energy. (b) A longitudinal optical phonon generates a local electric dipole $\boldsymbol{P}(\boldsymbol{R})$, which produces a local charge $\rho(\boldsymbol{R}) = -e^*\mathrm{div}\boldsymbol{P}(\boldsymbol{R})$.

Thus, an electron at $\boldsymbol{r}_{\mathrm{e}}$ feels the deformation potential at the same site: short-range-type interaction.

$$V_{\mathrm{int}}(\boldsymbol{r}_{\mathrm{e}} - \boldsymbol{R}) = -E_{\mathrm{d}}^{\mathrm{c}}\,\delta(\boldsymbol{r}_{\mathrm{e}} - \boldsymbol{R}), \qquad \text{hence } v_{\mathrm{int}}(\boldsymbol{k}) = -\frac{i}{\sqrt{N}} E_{\mathrm{d}}^{\mathrm{c}}. \tag{2.29}$$

Then, we finally obtain the interaction Hamiltonian by LA modes as

$$V_{\mathrm{eL}}(\boldsymbol{r}_{\mathrm{e}}) = -i\frac{1}{\sqrt{NM}}\sum_{\boldsymbol{k}} E_{\mathrm{d}}^{\mathrm{c}}\, k\, q_{\mathrm{LA},\boldsymbol{k}} \exp(i\,\boldsymbol{k}\cdot\boldsymbol{r}_{\mathrm{e}}). \tag{2.30}$$

Here, M is the total mass of a unit cell.

(2) Longitudinal optical (LO) mode

The longitudinal mode of optical phonons in ionic materials, such as NaCl, ZnO, GaAs, ..., generates local electric dipole moments, because a cation and an anion in a unit cell move oppositely with each other (Fig. 2.6b): The generating local electric dipole moment is the origin of the Fröhlich type interaction [4].

Let us consider a simple ionic material where a unit cell contains a cation (mass M_c) and an anion (mass M_a), and concentrate $\boldsymbol{k} = (k_x, 0, 0)$ LO mode, $q_{\mathrm{LA},\boldsymbol{k}}$, for a while. In the long wave limit $k_x \sim 0$, a cation and an anion in a cell at $\boldsymbol{R} = \boldsymbol{T}_l$ displace oppositely in proportion to

$$\Delta R^x_{l,\mathrm{c}} \propto M_\mathrm{a}/(M_\mathrm{c} + M_\mathrm{a}) \quad \text{and} \quad \Delta R^x_{l,\mathrm{a}} \propto M_\mathrm{c}/(M_\mathrm{c} + M_\mathrm{a}), \qquad (2.31)$$

respectively. This is because the center of mass does not move for $\boldsymbol{k} = 0$.

Then the coefficients in the displacements (1.38) have a relation,

$$|\Delta R^x_{l,\mathrm{c}}| : |\Delta R^x_{l,\mathrm{a}}| = \left|\frac{1}{\sqrt{M_\mathrm{c}}} e^x_{\mathrm{LO,c},\boldsymbol{k}}\right| : \left|\frac{1}{\sqrt{M_\mathrm{a}}} e^x_{\mathrm{LO,a},\boldsymbol{k}}\right| = \frac{M_\mathrm{a}}{M_\mathrm{c} + M_\mathrm{a}} : \frac{M_\mathrm{c}}{M_\mathrm{c} + M_\mathrm{a}}. \qquad (2.32)$$

Using Eq. (1.35a), ${e^x_{\mathrm{LO,c},\boldsymbol{k}}}^2 + {e^x_{\mathrm{LO,a},\boldsymbol{k}}}^2 = 1$ and $e^x_{\mathrm{LO,c},\boldsymbol{k}} // -e^x_{\mathrm{LO,a},\boldsymbol{k}}$, the induced relative ionic displacement caused by $q_{\mathrm{LA},\boldsymbol{k}}$ mode turns out to be

$$\Delta R^x_l \equiv \Delta R^x_{l,\mathrm{c}} - \Delta R^x_{l,\mathrm{a}} = \frac{1}{\sqrt{N\mu}} q_{\mathrm{LO},\boldsymbol{k}} \exp[i\, \boldsymbol{k} \cdot \boldsymbol{R}]. \qquad (2.33)$$

Here $\mu = (M_\mathrm{c}^{-1} + M_\mathrm{a}^{-1})^{-1}$ is the reduced mass. This displacement generates a local electric dipole moment $(e^*\Delta R^x_l, 0, 0)$. Here e^* is the effective charge of the ions. The result can be generalized for arbitral wave vector $\boldsymbol{k}$. An LO mode generates a local electric dipole density $\boldsymbol{p}(\boldsymbol{R}) = (N/V)e^*\Delta(\boldsymbol{R})$ parallel to $\boldsymbol{k}$ (N the number of unit cells, V the total volume).

Remember that dipoles in a uniform distribution are canceled with each other. Only its spatial derivative generates a net charge,

$$\rho(\boldsymbol{R}) = -\mathrm{div}\, \boldsymbol{p}(\boldsymbol{R}) = -\left(\frac{\partial}{\partial R_x} + \frac{\partial}{\partial R_y} + \frac{\partial}{\partial R_z}\right)\boldsymbol{p}(\boldsymbol{R})$$

$$= -i\, e^*\left(\sqrt{N/\mu}/V\right) |\boldsymbol{k}|\, q_{\mathrm{LO},\boldsymbol{k}} \exp[i\, \boldsymbol{k}\cdot\boldsymbol{R}]. \quad (2.34)$$

Then an electron at $\boldsymbol{r}_\mathrm{e}$ (far from $\boldsymbol{R}$) feels the Coulomb potential generated by these net charge $D_{\mathrm{LO}}(\boldsymbol{R}) \equiv \rho(\boldsymbol{R})$ with an interaction kernel

$$V_{\mathrm{int}}(\boldsymbol{r}_\mathrm{e} - \boldsymbol{R}) = \frac{e}{4\pi\epsilon\varepsilon_0}\frac{1}{|\boldsymbol{r}_\mathrm{e} - \boldsymbol{R}|}. \quad (2.35)$$

Here, ϵ is the relative effective dielectric constant of the material (to ε_0 in vacuum), whose details will be given below. Since the Fourier component of $V_{\mathrm{int}}(\boldsymbol{r})$ is

$$v_{\mathrm{int}}(\boldsymbol{k}) = \frac{e}{4\pi\epsilon\varepsilon_0}\frac{1}{|\boldsymbol{k}|^2}, \quad (2.36)$$

we obtain the interaction Hamiltonian for all LO modes by summing the contribution from all $\boldsymbol{k}$ in Eq. (2.24),

$$V_{\mathrm{int}}(\boldsymbol{r}_\mathrm{e}) = \sum_{\boldsymbol{k}} v_{\mathrm{int}}(\boldsymbol{k})\, d_{\sigma,\boldsymbol{k}} \exp[i\, \boldsymbol{k}\cdot\boldsymbol{r}_\mathrm{e}],$$

$$= -i\sum_{\boldsymbol{k}}\left[\frac{e^2\hbar\omega_{\mathrm{LO}}}{2V\varepsilon_0}\left(\frac{1}{\epsilon_\infty} - \frac{1}{\epsilon_\mathrm{s}}\right)\right]^{1/2}\frac{1}{|\boldsymbol{k}|}\, q_{\mathrm{LO},\boldsymbol{k}} \exp[i\, \boldsymbol{k}\cdot\boldsymbol{r}_\mathrm{e}], \quad (2.37)$$

Here V is the total volume, ϵ_s and ϵ_∞ are the static and high frequency relative dielectric constant of a material, respectively. The last form of Eq. (2.37) is derived after the Lyddane–Sachs–Teller relation with e^* and ω_{LO}, the optical phonon frequency [2, 4, 5].

The form of dielectric constants in Eq. (2.37) can be explained as follows. Consider a situation where an extra point charge is introduced at the origin in a dielectric material, the induced Coulomb field around the charge is given by $1/4\pi\epsilon_\mathrm{s}\varepsilon_0 r$. Here ϵ_s is the static dielectric constant that represents the induced dielectric fields which come from both the phonon filed and valence electrons: The magnitude of latter is evaluated as $1/4\pi\epsilon_s\varepsilon_0 r$, since in the high frequency limit, ions cannot follow the electric radiation field. Thus, the strength induced by the phonon field alone is proportional to $1/4\pi\epsilon_\infty\varepsilon_0 r$ - $1/4\pi\epsilon_s\varepsilon_0 r$. Then the effective constant is given by $\epsilon^{-1} = (\epsilon_\infty{}^{-1} - \epsilon_s{}^{-1})^{-1}$.

To summarize the results up to here, the effective Hamiltonian of a conduction electron and phonons in a material is

$$h_{\mathrm{e}}^{\mathrm{eff}} + V_{\mathrm{eL}} + H_{\mathrm{L}},$$

$$= \frac{\boldsymbol{p}_{\mathrm{e}}^2}{2m_{\mathrm{e}}^*} + \sum_{\sigma,\boldsymbol{k}} A_{\sigma,\boldsymbol{k}}^{\mathrm{c}}\, q_{\sigma,\boldsymbol{k}} \exp(i\,\boldsymbol{k}\cdot\boldsymbol{r}_{\mathrm{e}}) + \frac{1}{2}\sum_{\sigma,\boldsymbol{k}} (p_{\sigma,\boldsymbol{k}}^* p_{\sigma,\boldsymbol{k}} + \omega_{\sigma,\boldsymbol{k}}^2 q_{\sigma,\boldsymbol{k}}^* q_{\sigma,\boldsymbol{k}}). \tag{2.38a}$$

In the second quantization form for a conduction electron and phonons,

$$= \sum_{\boldsymbol{k}_{\mathrm{e}}} \frac{\hbar^2 k_{\mathrm{e}}^2}{2m_{\mathrm{e}}^*} a_{\mathrm{c},\boldsymbol{k}_{\mathrm{e}}}^{+} a_{\mathrm{c},\boldsymbol{k}_{\mathrm{e}}} + \sum_{\boldsymbol{k}_{\mathrm{e}}}\sum_{\sigma,\boldsymbol{k}} \sqrt{\frac{\hbar}{2\omega_{\sigma,\boldsymbol{k}}}} A_{\sigma,\boldsymbol{k}}^{\mathrm{c}} \left(b_{\sigma,\boldsymbol{k}} + b_{\sigma,-\boldsymbol{k}}^{+}\right) a_{\mathrm{c},\boldsymbol{k}_{\mathrm{e}}+\boldsymbol{k}}^{+} a_{\mathrm{c},\boldsymbol{k}_{\mathrm{e}}}$$
$$+ \sum_{\sigma,\boldsymbol{k}} \hbar\omega_{\sigma,\boldsymbol{k}} \left(b_{\sigma,\boldsymbol{k}}^{+} b_{\sigma,\boldsymbol{k}} + \frac{1}{2}\right). \tag{2.38b}$$

Summation on σ should be done over relevant phonon modes.

LA mode: deformation potential-type $\omega_{\mathrm{LA},\boldsymbol{k}} = uk$, u: sound velocity

$$V_{\mathrm{eL}} = \sum_{\boldsymbol{k}} A_{\mathrm{LA},\boldsymbol{k}}^{\mathrm{c}}\, q_{\mathrm{LA},\boldsymbol{k}} (i\,\boldsymbol{k}\cdot\boldsymbol{r}_{\mathrm{e}}) = -i\frac{1}{\sqrt{NM}} \sum_{\boldsymbol{k}} E_{\mathrm{d}}^{\mathrm{c}}\, k\, q_{\mathrm{LA},\boldsymbol{k}} \exp(i\,\boldsymbol{k}\cdot\boldsymbol{r}_{\mathrm{e}})$$
$$= -iE_{\mathrm{d}}^{\mathrm{c}} \sqrt{\frac{\hbar}{2NMu}} \sum_{\boldsymbol{k}} k^{1/2} \left(b_{\mathrm{LA},\boldsymbol{k}} + b_{\mathrm{LA},-\boldsymbol{k}}^{+}\right) \exp(i\,\boldsymbol{k}\cdot\boldsymbol{r}_{\mathrm{e}}). \tag{2.39a}$$

By using the creation and annihilation operators for a conduction electron,

$$V_{\mathrm{eL}} = -iE_{\mathrm{d}}^{\mathrm{c}} \sqrt{\frac{\hbar}{2NMu}} \sum_{\boldsymbol{k}_{\mathrm{e}}} \sum_{\boldsymbol{k}} k^{1/2} \left(b_{\mathrm{LA},\boldsymbol{k}} + b_{\mathrm{LA},-\boldsymbol{k}}^{+}\right) a_{\mathrm{c},\boldsymbol{k}_{\mathrm{e}}+\boldsymbol{k}}^{+}\, a_{\mathrm{c},\boldsymbol{k}_{\mathrm{e}}}. \tag{2.39b}$$

LO mode: Fröhlich interaction $\omega_{\mathrm{LO},\boldsymbol{k}} = \omega_{\mathrm{LO}}$: constant

$$V_{\mathrm{eL}} = \sum_{\boldsymbol{k}} A^{\mathrm{c}}_{\mathrm{LO},\boldsymbol{k}}\, q_{\mathrm{LO},\boldsymbol{k}}(i\,\boldsymbol{k}\cdot\boldsymbol{r}_{\mathrm{e}}) = -i\left[\frac{e^2\omega^2_{\mathrm{LO}}}{V\epsilon\varepsilon_0}\right]^{1/2}\sum_{\boldsymbol{k}}\frac{1}{|k|}q_{\mathrm{LO},\boldsymbol{k}}\exp[i\,\boldsymbol{k}\cdot\boldsymbol{r}_{\mathrm{e}}],$$

$$= -i\left[\frac{e^2\hbar\omega_{\mathrm{LO}}}{2V\epsilon\varepsilon_0}\right]^{1/2}\sum_{\boldsymbol{k}}\frac{1}{|k|}\left(b_{\mathrm{LO},\boldsymbol{k}}+b^{+}_{\mathrm{LO},-\boldsymbol{k}}\right)\exp(i\,\boldsymbol{k}\cdot\boldsymbol{r}_{\mathrm{e}}). \qquad (2.40a)$$

By using the creation and annihilation operators for a conduction electron,

$$V_{\mathrm{eL}} = -i\left[\frac{e^2\hbar\omega_{\mathrm{LO}}}{2V\epsilon\varepsilon_0}\right]^{1/2}\sum_{\boldsymbol{k}_{\mathrm{e}}}\sum_{\boldsymbol{k}}\frac{1}{|k|}\left(b_{\mathrm{LO},\boldsymbol{k}}+b^{+}_{\mathrm{LO},-\boldsymbol{k}}\right)a^{+}_{\mathrm{c},\boldsymbol{k}_{\mathrm{e}}+\boldsymbol{k}}a_{\mathrm{c},\boldsymbol{k}_{\mathrm{e}}}. \qquad (2.40b)$$

In nonpolar semiconductors, such as silicon, the optical phonon modes do not induce any electric polarization, but they induce short-range potential for a conduction electron. The coefficient can be derived similarly as for the acoustic phonons. See details in Ref. [6]. For elementary methods to derive the coefficients of the electron–lattice interaction, there are two basic methods: frozen phonon method and density functional perturbation theory [7, 8].

2.2.4 A Valence Hole

How about a hole in the valence band? This is an (N_{e} − 1) electron problem under scheme (B) in Chapter 1. In principle, this problem can be treated by calculating the adiabatic potentials $U_q^{N_{\mathrm{e}}-1}(\{\Delta\boldsymbol{R}\})$ for the q-th state of (N_{e} − 1) electrons as a function of the atomic (ionic) displacements $\{\Delta\boldsymbol{R}\}$. As it has been done on a conduction electron just above, we can extend a similar discussion, but with sufficient carefulness. In order to make physics clear, we focus on the following situation (Fig. 2.2c).

(a) Electronic state

(N_{e} − 1) electrons occupy the lowest possible one-electron states, except the q-th one-electron state $\psi_{\mathrm{v},q,\{\Delta\boldsymbol{R}\}}(\boldsymbol{r})$ in the valence band whose one-electron energy is $\varepsilon_{\mathrm{v},q}(\{\Delta\boldsymbol{R}\})$.

The unoccupied one-electron state $\psi_{\mathrm{v},q,\{\Delta\boldsymbol{R}\}}(\boldsymbol{r})$ is arbitral but can be expressed either by a linear combination of Bloch states $\{\psi_{\mathrm{v},\boldsymbol{k}_h}(\boldsymbol{r})\}$ or by Wannier states $\{\phi_{\mathrm{v},\boldsymbol{T}_l}(\boldsymbol{r})\}$ of the valence band. Thus, the present (N_e – 1) electron state can be expressed in terms of

$$\prod_{\boldsymbol{k}\neq\boldsymbol{k}_h} a^+_{\mathrm{v},\boldsymbol{k}}\,|0\rangle = a_{\mathrm{v},\boldsymbol{k}_h}|\mathrm{G0}\rangle \leftrightarrow \psi_{\mathrm{v},\boldsymbol{k}_h}(\boldsymbol{r}) \text{ is empty } \leftrightarrow \text{a hole in } \psi_{\mathrm{v},\boldsymbol{k}_h}(\boldsymbol{r}). \tag{2.41a}$$

or

$$\prod_{\boldsymbol{T}_m\neq\boldsymbol{T}_l} a^+_{\mathrm{v},\boldsymbol{T}_m}\,|0\rangle = a_{\mathrm{v},\boldsymbol{T}_l}|\mathrm{G0}\rangle \leftrightarrow \phi_{\mathrm{v},\boldsymbol{T}_l}(\boldsymbol{r}) \text{ is empty} \leftrightarrow \text{a hole in } \phi_{\mathrm{v},\boldsymbol{T}_l}(\boldsymbol{r}). \tag{2.41b}$$

Here $a_{\mathrm{v},\boldsymbol{k}}$ and $a_{\mathrm{v},\boldsymbol{T}_l}$ are the electron annihilation operator, i.e. hole creation operator, for a Bloch orbital $\psi_{\mathrm{v},\boldsymbol{k}}(\boldsymbol{r})$ with a wave vector $\boldsymbol{k}$ and for a Wannier orbital $\phi_{\mathrm{v},\boldsymbol{T}_l}(\boldsymbol{r})$ at unit cell $\boldsymbol{T}_l$, respectively. Both operators are related with a Fourier transformation:

$$a_{\mathrm{v},\boldsymbol{T}_l} = \frac{1}{\sqrt{N}}\sum_{\boldsymbol{k}} \exp(i\boldsymbol{k}\cdot\boldsymbol{T}_l)\, a_{\mathrm{v},\boldsymbol{k}}. \tag{2.42}$$

(b) Adiabatic potentials

In the Hartree–Fock approximation, Koopmans' theorem states that

$$U_q^{N_\mathrm{e}}(\{\Delta\boldsymbol{R}\}) = \varepsilon_{\mathrm{v},q}(\{\Delta\boldsymbol{R}\}) + U_\mathrm{g}^{N_\mathrm{e}-1}(\{\Delta\boldsymbol{R}\}). \tag{2.43}$$

Then,

$$U_q^{N_\mathrm{e}-1}(\{\Delta\boldsymbol{R}\}) = -\varepsilon_{\mathrm{v},q}(\{\Delta\boldsymbol{R}\}) + U_\mathrm{g}^{N_\mathrm{e}}(\{\Delta\boldsymbol{R}\}). \tag{2.44}$$

Here, $U_\mathrm{g}^{N_\mathrm{e}}(\{\Delta\boldsymbol{R}\})$ is the adiabatic potential for the ground state of N_e electron. As has been discussed, $U_\mathrm{g}^{N_\mathrm{e}}(\{\Delta\boldsymbol{R}\})$ includes all $\varepsilon_{\mathrm{v},l}(\{\Delta\boldsymbol{R}\})$ dependence on the atomic displacements $\{\Delta\boldsymbol{R}\}$ and the interaction among ions, resulting in the so-called lattice elastic energy. Then the adiabatic potential of the present (N_e – 1) electron state is expressed as

$$U_q^{N_e-1}(\{\Delta \boldsymbol{R}\}) = -\varepsilon_q(\{\Delta \boldsymbol{R}\}) + \frac{1}{2}\sum_{\sigma,\boldsymbol{k}} \omega_{\sigma,\boldsymbol{k}}^2 q_{\sigma,\boldsymbol{k}}^2, \tag{2.45}$$

which enables us to treat the present "(N_e – 1) electron" system as a hole in the valence band + the lattice vibration.

$$h_h^{eff} + V_{hL} + H_L, \tag{2.46}$$

(c) Effective Hamiltonian for hole part: h_h

The wave function is described by a linear combination of Wannier states $\phi_{v,\boldsymbol{T}_l}$, where $c_{\boldsymbol{T}_l}$ is a slowly varying function of the lattice point $\boldsymbol{T}_l$.

$$\psi_v = \sum_{\boldsymbol{T}_l} c_{\boldsymbol{T}_l}\, \phi_{v,\boldsymbol{T}_l} \tag{2.47}$$

Replacing the discrete index $\boldsymbol{T}_l$ by a continuous variable $\boldsymbol{r}_h$ for a hole, the coefficient $c_{\boldsymbol{T}_l}$ itself can be considered as a wave function of $\boldsymbol{r}_h$,

$$c_{\boldsymbol{T}_l} \quad \to \quad \psi_v(\boldsymbol{r}_h) \tag{2.48}$$

Procedure to reduce the effective Hamiltonian for a valence hole with the envelope function approximation can be derived similarly as in Section 2.2.2, and the result is summarized as

$$h_h^{eff} = \frac{\boldsymbol{p}_h^2}{2m_h^*}. \tag{2.49}$$

Here m_h^* is the effective mass of a valence hole, defined by

$$\frac{1}{m_h^*} = -\frac{1}{\hbar^2}\frac{\partial^2 \varepsilon_v(\boldsymbol{k})}{\partial \boldsymbol{k}^2} \tag{2.50}$$

where $\varepsilon_v(\boldsymbol{k})$ is the energy dispersion of the valence band and the second derivative is taken at the top of the band. If crystals are not isotropic, the effective mass has a tensor quantity. In almost covalent semiconductors, near the top of the valence band, Bloch states are degenerated and split by the spin-orbit coupling.

(d) Effective Hamiltonian for interaction part: V_{hL}

Since a hole has a positive charge opposite to an electron, we can derive similar expressions for the hole–lattice interactions.

For the interaction with optical phonons, we have

$$A^{\mathrm{v}}_{\mathrm{op},\boldsymbol{k}} = -A^{\mathrm{c}}_{\mathrm{op},\boldsymbol{k}}, \tag{2.51}$$

As for the interaction with the acoustic mode, $A^{\mathrm{v}}_{\mathrm{ac},\boldsymbol{k}}$ is not related with $A^{\mathrm{c}}_{\mathrm{ac},\boldsymbol{k}}$, they are independent quantity. In conclusion, the effective interaction for a hole in the valence band interacting with lattice vibrations is given by a set of similar equations with replacing the suffix from c to v in Eqs. (2.38–2.40).

As a summary, the effective Hamiltonian of excited state: a conduction electron and a valence hole plus the lattice vibrations are

$$h^{\mathrm{eff}}_{\mathrm{e}} + V_{\mathrm{eL}} + h^{\mathrm{eff}}_{\mathrm{h}} + V_{\mathrm{hL}} + V_{\mathrm{eh}} + H_{\mathrm{L}},$$

$$= \frac{\boldsymbol{p}_{\mathrm{e}}^2}{2m_{\mathrm{e}}^*} + \sum_{\sigma,\boldsymbol{k}} A^{\mathrm{c}}_{\sigma,\boldsymbol{k}}\, q_{\sigma,\boldsymbol{k}} \exp(i\,\boldsymbol{k}\cdot\boldsymbol{r}_{\mathrm{e}}) + \frac{\boldsymbol{p}_{\mathrm{e}}^2}{2m_{\mathrm{e}}^*} + \sum_{\sigma,\boldsymbol{k}} A^{\mathrm{v}}_{\sigma,\boldsymbol{k}}\, q_{\sigma,\boldsymbol{k}} \exp(i\,\boldsymbol{k}\cdot\boldsymbol{r}_{\mathrm{h}})$$

$$-\frac{e^2}{4\pi\epsilon_\infty\varepsilon_0}\frac{1}{|\boldsymbol{r}_{\mathrm{e}}-\boldsymbol{r}_{\mathrm{h}}|} + \frac{1}{2}\sum_{\sigma,\boldsymbol{k}} (p^*_{\sigma,\boldsymbol{k}} p_{\sigma,\boldsymbol{k}} + \omega^2_{\sigma,\boldsymbol{k}} q^*_{\sigma,\boldsymbol{k}} q_{\sigma,\boldsymbol{k}}). \tag{2.52}$$

Here we add V_{eh}, the Coulomb attraction between an electron and a hole. Note that almost all equations in this chapter are valid within the effective mass approximation. The parameter values on the electron–lattice interactions in various semiconductors, such as deformation potential, dielectric constant, effective mass, stiffness constant, ..., are found in handbooks on semiconductors, for example, Refs [9, 10].

In final, it would be worthy here to remark on the spin of a hole. If we subtract an electron from a Bloch state with up spin, $\psi_{\mathrm{v},\boldsymbol{k1},\uparrow}(\boldsymbol{r})$, in the valence band

$$\psi_{\mathrm{v},\boldsymbol{k1},\uparrow}(\boldsymbol{r}) \text{ is empty} \;\leftrightarrow\; a_{\mathrm{v},\boldsymbol{k1},\uparrow}|\mathrm{G0}\rangle = a_{\mathrm{v},\boldsymbol{k1},\uparrow} \prod_{\boldsymbol{k}=\mathrm{occupied},\alpha=\uparrow,\downarrow} a^+_{\mathrm{v},\boldsymbol{k},\alpha}|0\rangle$$

$$= \prod_{\boldsymbol{k}\neq\boldsymbol{k1},\alpha=\uparrow,\downarrow} a^+_{\mathrm{v},\boldsymbol{k},\alpha} a^+_{\mathrm{v},\boldsymbol{k1},\downarrow}|0\rangle. \tag{2.53}$$

Thus, all the up and down spin for each $\boldsymbol{k}$ state cancel each other except $\boldsymbol{k}1$ state, then total spin of this state is a spin down.

$$\psi_{\mathrm{v},\boldsymbol{k1},\uparrow}(\boldsymbol{r}) \text{ is empty} \leftrightarrow a_{\mathrm{v},\boldsymbol{k1},\uparrow}|\mathrm{G0}\rangle \leftrightarrow \text{a hole with momentum } -\hbar\boldsymbol{k}_1, \downarrow \text{spin}$$

$$\psi_{\mathrm{v},\boldsymbol{k1},\downarrow}(\boldsymbol{r}) \text{ is empty} \leftrightarrow a_{\mathrm{v},\boldsymbol{k1},\downarrow}|\mathrm{G0}\rangle \leftrightarrow \text{a hole with momentum } -\hbar\boldsymbol{k}_1, \uparrow \text{spin}$$

2.3 Carrier Scattering

We have seen that when an extra electron (hole) is introduced in the conduction (valence) band, it interacts with the lattice vibrations. Hereafter, using this effective Hamiltonian (2.38b), we will investigate various phenomena induced by the carrier–phonon interaction.

$$h_{\mathrm{e}}^{\mathrm{eff}} + V_{\mathrm{eL}} + H_{\mathrm{L}}$$

$$= \sum_{\boldsymbol{k}_{\mathrm{e}}} \frac{\hbar^2 k_{\mathrm{e}}^2}{2m_{\mathrm{e}}^*} a_{\mathrm{c},\boldsymbol{k}_{\mathrm{e}}}^+ a_{\mathrm{c},\boldsymbol{k}_{\mathrm{e}}} + \sum_{\boldsymbol{k}_{\mathrm{e}}} \sum_{\sigma,\boldsymbol{k}} \sqrt{\frac{\hbar}{2NM\omega_{\sigma,\boldsymbol{k}}}} A_{\sigma,\boldsymbol{k}}^{\mathrm{c}} \left(b_{\sigma,\boldsymbol{k}} + b_{\sigma,-\boldsymbol{k}}^+\right) a_{\mathrm{c},\boldsymbol{k}_{\mathrm{e}}+\boldsymbol{k}}^+ a_{\mathrm{c},\boldsymbol{k}_{\mathrm{e}}}$$

$$+ \sum_{\sigma,\boldsymbol{k}} \hbar\omega_{\sigma,\boldsymbol{k}} \left(b_{\sigma,\boldsymbol{k}}^+ b_{\sigma,\boldsymbol{k}} + \frac{1}{2}\right).$$

Let us treat V_{eL} as a perturbation. The eigenstates of $h_{\mathrm{e}} + H_{\mathrm{L}}$ are direct products of electron part $\psi_{\mathrm{c},k_{\mathrm{e}}}$ and phonon part $\{n_{\sigma,\boldsymbol{k}}\}$

$$\left|\psi_{\mathrm{c},\boldsymbol{k}_{\mathrm{e}}}; \{n_{\sigma,\boldsymbol{q}}\}\right\rangle = a_{\mathrm{c},\boldsymbol{k}_{\mathrm{e}}}^+ \prod_{\sigma,\boldsymbol{k}} \frac{1}{\sqrt{n_{\sigma,\boldsymbol{k}}}} (b_{\sigma,\boldsymbol{k}}^+)^{n_{\sigma,\boldsymbol{k}}} |\boldsymbol{0}\rangle. \tag{2.54a}$$

Here $|\boldsymbol{0}\rangle$ is the electron–phonon vacuum. The corresponding eigenenergy is

$$E(\boldsymbol{k}_{\mathrm{e}}; \{n_{\sigma,\boldsymbol{k}}\}) = \varepsilon_{\mathrm{c}}(\boldsymbol{k}_{\mathrm{e}}) + \sum_{\sigma,\boldsymbol{k}} \hbar\omega_{\sigma,\boldsymbol{k}} \left(n_{\sigma,\boldsymbol{k}} + \frac{1}{2}\right). \quad n_{\sigma,\boldsymbol{k}} = 0, 1, 2, \ldots \tag{2.54b}$$

This state describes a conduction electron in $\psi_{\mathrm{c},k_{\mathrm{e}}}(r_{\mathrm{e}})$ of a wave vector $\boldsymbol{k}_{\mathrm{e}}$ and energy $\varepsilon_{\mathrm{c}}(\boldsymbol{k}_{\mathrm{e}})$ moves in a semiconductor with a set of phonon states $\{n_{\sigma,\boldsymbol{k}}\}$: σ the phonon mode index and the wave vector $\boldsymbol{k}$. Since the system has a translational symmetry, the total momentum is a constant of motion.

$$\boldsymbol{P}_{\text{tot}} \equiv \hbar \boldsymbol{k}_{\text{e}} + \sum_{\sigma,\boldsymbol{k}} \hbar \boldsymbol{k} b^{+}_{\sigma,\boldsymbol{k}} b_{\sigma,\boldsymbol{k}}. \tag{2.55}$$

In other words, $\boldsymbol{P}_{\text{tot}}$ commutes with H_{eff}: $\boldsymbol{P}_{\text{tot}} H_{\text{eff}} = H_{\text{eff}} \boldsymbol{P}_{\text{tot}}$.

When the interaction V_{eL} is switched on, the state $|\psi_{\text{c},\boldsymbol{k}_{\text{e}}}; \{n_{\sigma,\boldsymbol{k}}\}\rangle$ is perturbed and no longer a steady state. The transition rate $r_{kk'}$ to a final state $|\psi_{\text{c},\boldsymbol{k}'}; \{n_{\sigma,\boldsymbol{q}'}\rangle$ is given by the Fermi's golden rule

$$r_{\boldsymbol{k}_{\text{e}}\boldsymbol{k}_{\text{e}}'} = \frac{2\pi}{\hbar} \left| \langle \psi_{\text{c},\boldsymbol{k}_{\text{e}}\sigma,\boldsymbol{k}'}; \{n_{\sigma,\boldsymbol{k}'}\} | V_{\text{eL}} | \psi_{\text{c},\boldsymbol{k}_{\text{e}}}; \{n_{\sigma,\boldsymbol{k}}\} \rangle \right|^2$$

$$\times\, \delta\left(\varepsilon(\boldsymbol{k}_{\text{e}}) + \sum_{\sigma,\boldsymbol{k}} \hbar\omega_{\sigma,\boldsymbol{k}} \left(n_{\sigma,\boldsymbol{k}} + \frac{1}{2} \right) - \varepsilon(\boldsymbol{k}_{\text{e}}') - \sum_{\sigma,\boldsymbol{k}} \hbar\omega_{\sigma,\boldsymbol{k}} \left(n_{\sigma,\boldsymbol{k}'} + \frac{1}{2} \right) \right). \tag{2.56}$$

The matrix element contains $(b_{\sigma,\boldsymbol{k}} + b^{+}_{\sigma,-\boldsymbol{k}}) a^{+}_{\text{c},\boldsymbol{k}_{\text{e}}+\boldsymbol{k}} a_{\text{c},\boldsymbol{k}_{\text{e}}}$ operators, then there are two possible processes (Fig. 2.7) whose matrix element proportional to

one-phonon absorption

$$\langle n_{\sigma,\boldsymbol{k}} - 1 | b^{+}_{\sigma,-\boldsymbol{k}} | n_{\sigma,\boldsymbol{k}} \rangle = \sqrt{n_{\sigma,\boldsymbol{k}}}, \tag{2.57a}$$

one-phonon emission

$$\langle n_{\sigma,\boldsymbol{k}} + 1 | b^{+}_{\sigma,-\boldsymbol{k}} | n_{\sigma,\boldsymbol{k}} \rangle = \sqrt{n_{\sigma,\boldsymbol{k}} + 1}, \tag{2.57b}$$

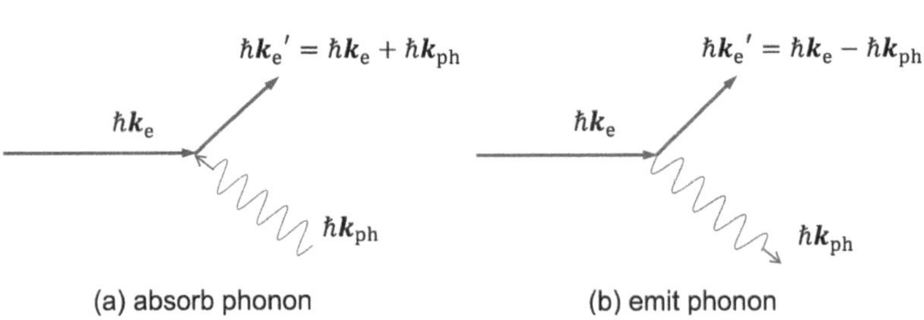

Figure 2.7 A conduction electron is scattered by a phonon (a) absorption and (b) emission processes.

Thus, both transition rates are proportional to the existing phonon number $n_{\sigma,\boldsymbol{k}}$ or $n_{\sigma,\boldsymbol{k}} + 1$; this enhancement is typical to Bose particles. These two processes obey the energy conservation,

$$\varepsilon(\boldsymbol{k}_{\mathrm{e}}{}') = \varepsilon(\boldsymbol{k}_{\mathrm{e}}) \pm \hbar\omega_{\sigma,\boldsymbol{k}}, \tag{2.58a}$$

and the momentum conservation

$$\hbar\boldsymbol{k}_{\mathrm{e}}{}' = \hbar\boldsymbol{k}_{\mathrm{e}} \pm \hbar\boldsymbol{k} \pm \hbar\boldsymbol{G}_{m}. \tag{2.58b}$$

$\boldsymbol{G}_m \neq \boldsymbol{0}$ process is called the Umklapp process, which rarely takes place for low-energy carriers. Figure 2.8 shows possible scattering processes by the interaction with (a) LA phonon and (b) LO phonon. Depending on the k-dependence of the e-phonon coupling constant $A^{\mathrm{c}}_{\sigma,\boldsymbol{k}}$ (Eqs. (2.37b, 2.38b)), the transition rate $r_{\boldsymbol{k}_{\mathrm{e}}\boldsymbol{k}_{\mathrm{e}'}}$ is proportional to

$$\left|A^{\mathrm{c}}_{\mathrm{LA},\boldsymbol{k}}\right|^2 \propto k, \quad \text{LA deformation potention type} \tag{2.59a}$$

$$\left|A^{\mathrm{c}}_{\mathrm{LO},\boldsymbol{k}}\right|^2 \propto k^{-2}. \quad \text{LO Fröhlich type} \tag{2.59b}$$

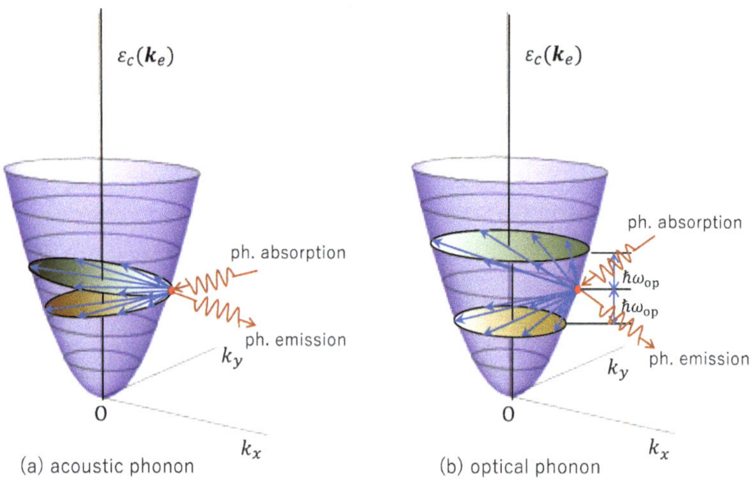

Figure 2.8 Illustration of possible scattering processes in two-dimensional $\boldsymbol{k}$-space by (a) acoustic phonon and (b) optical phonon. Color-density indicates proportionally the $\boldsymbol{k}$-dependence in the transition rate $r_{\boldsymbol{k}_{\mathrm{e}}\boldsymbol{k}_{\mathrm{e}'}}$.

The lifetime τ of a free electron is given by

$$\tau = \left[\sum_{k_{e'}} r_{k_e k_{e'}}(1 - f(\varepsilon(\boldsymbol{k}_e')))\right]^{-1}. \quad (2.60)$$

The factor 1 - $f(\varepsilon(\boldsymbol{k}_e'))$ is a result from the Pauli exclusion principle, when the system contains many electrons in the conduction band. Here $f(\varepsilon)$ is the Fermi distribution function.

$$f(\varepsilon) = \frac{1}{\exp\left[\frac{\varepsilon - \varepsilon_F}{k_B T}\right] - 1}, \quad (2.61)$$

where ε_F is the Fermi energy and k_B the Boltzmann constant.

As a summary, we find the mobility of a conduction electron,

$$\mu = e/m_e^* \tau, \quad (2.62)$$

which shows a temperature dependence, originating from the phonon scattering rate, which depends on the average phonon number

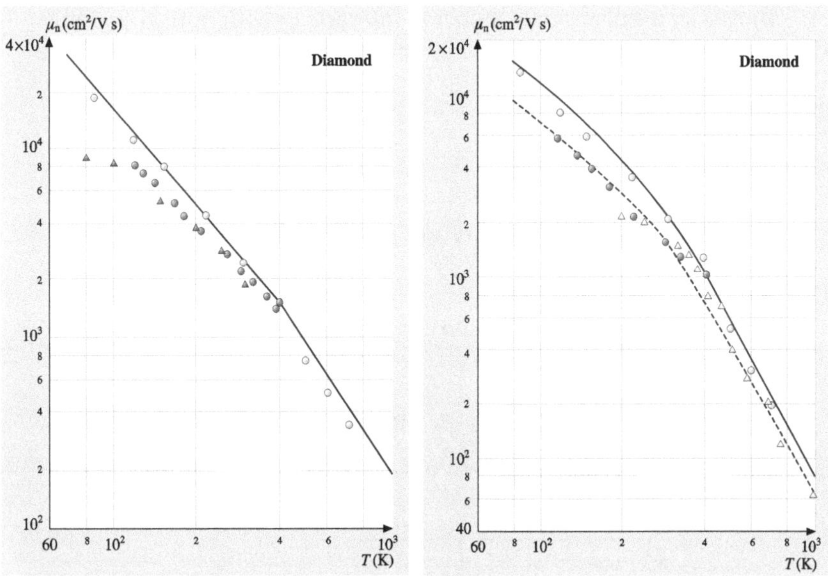

(a) electron (b) hole

Figure 2.9 The temperature dependence of the carrier mobility in diamond for (a) a conduction electron and (b) a valence hole. Reproduced with permission from Ref. [9].

$$\langle n_{\sigma,k}\rangle = \frac{1}{\exp\left[\hbar\omega_{\sigma,k}/k_{\mathrm{B}}T\right]-1}. \tag{2.63}$$

Figure 2.9 shows typical temperature dependence of a carrier mobility in Si [9]. Details can be found in typical textbooks on semiconductor physics.

2.4 Polaron: Carrier Dressed with Phonons

An extra electron (hole) in the conduction (valence) band thus interacts with the lattice vibrations. There arises a further situation: A conduction electron moves in a crystal emitting and absorbing phonons to form a phonon cloud around it. This is called a polaron, which has been intensively studied in the 1950s [11].

2.4.1 Fröhlich Polaron

Consider a carrier (conduction electron or valence hole) that moves in a polar semiconductor interacting with optical phonons via Fröhlich interaction. The situation is described by the effective Hamiltonian

$$h_{\mathrm{e}}^{\mathrm{eff}} + V_{\mathrm{eL}} + H_{\mathrm{L}},$$

$$= \frac{\boldsymbol{p}_{\mathrm{e}}^2}{2m_{\mathrm{e}}^*} + \sum_{\boldsymbol{k}}\left(V_{\boldsymbol{k}} b_{\mathrm{LO},\boldsymbol{k}} + V_{\boldsymbol{k}}^* b_{\mathrm{LO},-\boldsymbol{k}}^{+}\right) \mathrm{e}^{i\boldsymbol{q}\cdot\boldsymbol{r}_{\mathrm{e}}} + \sum_{\sigma,\boldsymbol{k}} \omega_{\mathrm{LO}}\left(b_{\mathrm{LO},\boldsymbol{k}}^{+} b_{\mathrm{LO},\boldsymbol{k}} + \frac{1}{2}\right).$$

$$V_{\boldsymbol{k}} = -\frac{i\hbar\omega_{\mathrm{LO}}}{|\boldsymbol{k}|}\left(\frac{\hbar}{2m\omega_{\mathrm{LO}}}\right)^{1/4}\left(\frac{4\pi\alpha}{V}\right)^{1/2}. \tag{2.64}$$

Here we have introduced a dimensionless (Fröhlich) polaron coupling constant

$$\alpha \equiv \frac{1}{2\hbar\omega_{\mathrm{LO}}\varepsilon_0}\left(\frac{1}{\epsilon_\infty} - \frac{1}{\epsilon_{\mathrm{s}}}\right)\frac{e^2}{\left[\frac{\hbar}{2m\omega_{\mathrm{LO}}}\right]^{1/2}}. \tag{2.65}$$

$\left[\frac{\hbar}{2m\omega_{\mathrm{LO}}}\right]^{1/2} \equiv r_p$ denotes the effective polaron radius; then α denotes the lattice relaxation energy divided by a phonon energy $\hbar\omega_{\mathrm{LO}}$.

Let us again treat V_{eL} as a perturbation. The eigenstates of $h_{\mathrm{e}} + H_{\mathrm{L}}$ are given by Eq. (2.54a). For simplicity, we omit the suffix c and op.

Zeroth order

$$|\boldsymbol{k}_{\mathrm{e}}; \{n_{\boldsymbol{k}} = 0\}\rangle, \qquad E\big(\boldsymbol{k}_{\mathrm{e}}; \{n_{\boldsymbol{q}} = 0\}\big) = \varepsilon_{\boldsymbol{k}_{\mathrm{e}}} = \frac{\hbar^2 {\boldsymbol{k}_{\mathrm{e}}}^2}{2m_{\mathrm{e}}^*}. \tag{2.66a}$$

First order

$$|\boldsymbol{k}_{\mathrm{e}}; \{n_{\boldsymbol{k}}\}\rangle + \sum_{\boldsymbol{k}} |\boldsymbol{k}_{\mathrm{e}} - \boldsymbol{k}; \{n_{\boldsymbol{k}} = 1\}\rangle \frac{\langle \boldsymbol{k}_{\mathrm{e}} - \boldsymbol{k}; \{n_{\boldsymbol{k}} = 1\}|V_{\mathrm{eL}}|\boldsymbol{k}_{\mathrm{e}}; \{n_{\boldsymbol{k}} = 0\}\rangle}{\varepsilon_{\boldsymbol{k}_{\mathrm{e}}} - (\varepsilon_{\boldsymbol{k}_{\mathrm{e}}-\boldsymbol{k}} + \hbar\omega_{\mathrm{op}})}$$

$$E(\boldsymbol{k}_{\mathrm{e}}; \{n_{\boldsymbol{k}} = 0\}) = \varepsilon_{\boldsymbol{k}_{\mathrm{e}}}, \tag{2.66b}$$

Second order (Fig. 2.10)

$$E(\boldsymbol{k}_{\mathrm{e}}; \{n_{\boldsymbol{k}}\}) = \varepsilon_{\boldsymbol{k}_{\mathrm{e}}} + \sum_{\boldsymbol{k}} \frac{|\langle \boldsymbol{k}_{\mathrm{e}} - \boldsymbol{k}; \{n_{\boldsymbol{k}} = 1\}|V_{\mathrm{eL}}|\boldsymbol{k}_{\mathrm{e}}; \{n_{\boldsymbol{k}} = 0\}\rangle|^2}{\varepsilon_{\boldsymbol{k}_{\mathrm{e}}} - (\varepsilon_{\boldsymbol{k}_{\mathrm{e}}-\boldsymbol{k}} + \hbar\omega_{\mathrm{op}})}, \tag{2.66c}$$

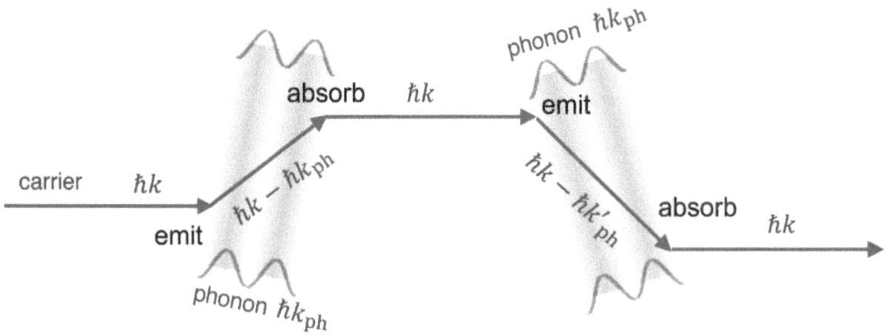

Figure 2.10 Symbolical illustration of a polaron: A carrier is dressed with self-induced phonon cloud by emitting and absorbing phonons.

Putting the $\boldsymbol{k}$-dependence of the Fröhlich interaction (2.38) into equations, we obtain the energy of a polaron.

$$E(\boldsymbol{k}_{\mathrm{e}}; \{n_{\boldsymbol{k}} = 0\}) = \frac{\hbar^2 {k_{\mathrm{e}}}^2}{2m_{\mathrm{e}}^*}\left(1 + \frac{\alpha}{6}\right). \tag{2.67}$$

The result indicates that the effective mass increases as

$$m_{\text{polaron}} = m_e^*(1 + \alpha/6), \tag{2.68}$$

due to accompanied phonon clouds.

We can further calculate the energy of a polaron for the higher order of the coupling constant α. We have seen that the interaction induces lattice shifts along each mode. Lie, Low, and Pines [12] tried a trial wave function for the intermediate coupling strength using a Unitary transformation

$$U = \exp\left[\sum_{\boldsymbol{k}} (b_{\boldsymbol{k}}^{+} f(\boldsymbol{k}) - b_{\boldsymbol{k}} f^{*}(\boldsymbol{k}))\right], \tag{2.69}$$

where $f(\boldsymbol{k})$ is a variable parameter to obtain appropriate displacements (details on the shift operator will be discussed in Appendix II).

$$U^{-1} b_{\boldsymbol{k}}^{+} U = b_{\boldsymbol{k}}^{+} + f^{*}(\boldsymbol{k}), \quad U^{-1} b_{\boldsymbol{k}} U = b_{\boldsymbol{k}} + f(\boldsymbol{k}). \tag{2.70}$$

Figure 2.11 shows the calculated results of various approximations. It can be seen that as the coupling constant α increases the dressed phonon density and the effective mass increases continuously [13].

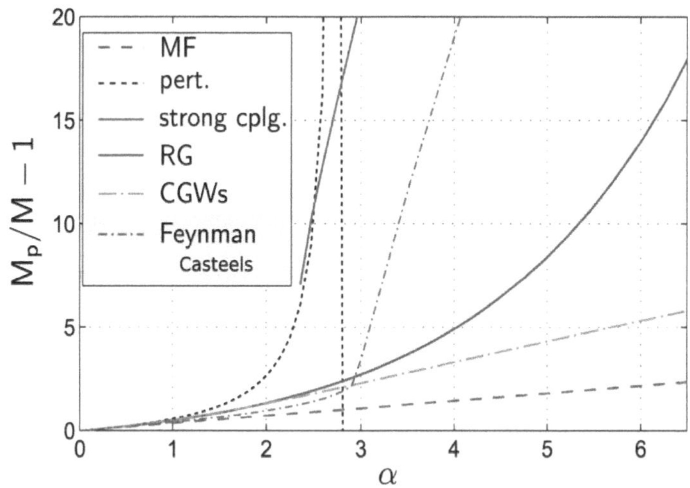

Figure 2.11 Polaron mass as a function of the electron–phonon coupling α. Reproduced from Ref. [13].

Feynman [14] applied a path-integral method to the polaron problem for entire range of the coupling constant α. The result also shows a continuous increase of the effective mass.

2.4.2 Polaron by Deformation Potential and Self-Trapping

We have seen above that when the strength of the carrier–lattice interaction is large, induced phonon clouds around the carrier becomes so dense resulting in an increase in the effective mass. In the strong coupling limit, the polaron could not move at all due to self-induced potential well: This phenomenon is called self-trapping (Fig. 2.12).

How about the short-range interaction by the LA mode via deformation potential interaction? Toyozawa [15] calculated the polaron state using a trial function similar to Eq. (2.66). He showed that the effective mass of the polaron is multivalued for intermediate coupling constant, resulting in a sudden increase of the effective mass, as shown in Fig. 2.13. This result is contrary to the interaction with LO modes, which shows continuous increase of the polaron mass. A self-trapping of a carrier is probable for the short-range interaction by the acoustic phonons [15–17]. Self-trapping of a carrier will be precisely discussed in Chapter 4.

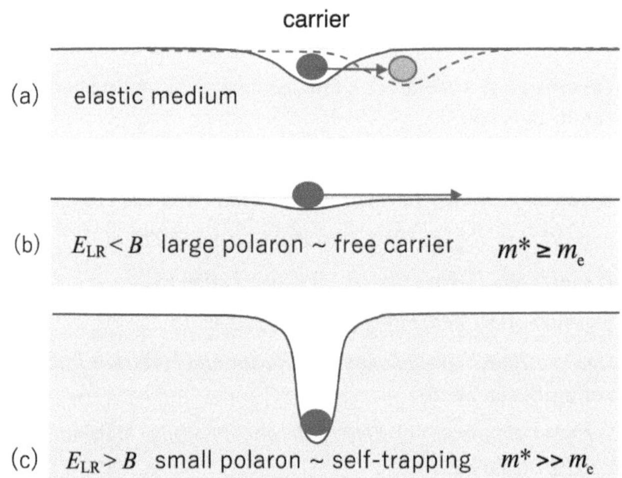

Figure 2.12 Illustration of (a) polaron, (b) large polaron, and (c) small polaron.

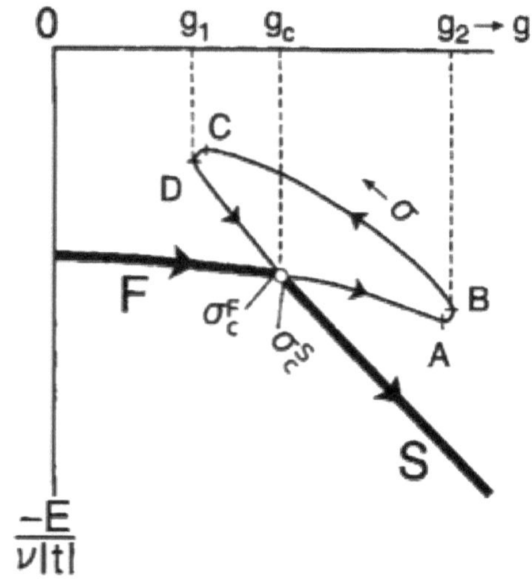

Figure 2.13 Plot of the polaron energy as a function of the coupling constant g of the deformation potential type interaction, where σ denotes the magnitude of the overlap of phonon states between nearest neighbors. Reproduced with permission from Ref. [15].

References

1. Shinozuka, Y., *Jpn. J. Appl. Phys.*, **32** (1993) 4560.
2. Ueta, M., Kanzaki, H., Kobayashi, K., Toyozawa, Y., and Hanamura, E., *Excitonic Processes in Condensed Matter* (Springer, 1986).
3. Toyozawa, Y, *Optical Processes in Solids* (Cambridge Univ. Press, Cambridge, 2003).
4. Fröhlich, H., *Theory of Dielectronics* (Oxford, 1949).
5. Lyddane, R., Sachs, R., Teller, E., *Phys. Rev.*, **59** (1941) 673.
6. Pötz, W. and Vogl,P., *Phys. Rev., B*, **24** (1981) 2025.
7. Frederiksen, T., *Phys Rev., B*, **75** (2007) 205413.
8. Baroni, S., *Rev. Mod. Phys.*, **73** (2001) 515.
9. *Springer Handbook of Condensed Matter and Materials Data* (Springer-Verlag, Berlin 2005).
10. *Springer Handbook of Electronic and Photonic Materials* (Springer-Verlag, Berlin 2006).
11. Fröhlich, H., Pelzer, H., and Zienau, S., *Phil. Mag.*, **41** (1950) 221.

12. Lee, T. D., Low, F. E., and Pines, D., *Phys. Rev.*, **90** (1953) 297.
13. Grusdt,F., Shchadilova, Y.E., Rubtsov, A.N., and Demler, E., *Scientific Reports*, **17** (2015), 5:12124.
14. Feynman, R. P., *Phys. Rev.*, **97** (1955) 660.
15. Toyozawa, Y., *Prog. Theor. Phys.*, **26** (1961) 29.
16. Kuper, C. G., and Whitfield, G. D., *Polarons and Excitons* (Oliver and Boyd, Edinburgh and London, 1963).
17. Song, K. S., and Williams, R. T., *Self-Trapped Excitons* (Springer, 1993).

Chapter 3

Configuration Coordinate Diagram

Electron–lattice interactions in materials induce various phenomena, which are called polaron effects in general. In order to discuss various polaron effects, it is helpful to introduce the configuration coordinate diagram (CCD). This chapter deals with CCD, which is useful especially for strong electron–lattice interaction appearing in localized systems. Here we start with the definition of CCD and the physical meaning. With the aid of CCD, we discuss the lattice relaxation dynamics, energy dissipation, photoabsorption, and photoemission.

3.1 Adiabatic Instability

The configuration coordinate diagram (CCD) is a useful tool for discussing various electronic-atomic processes induced by strong electron–lattice interactions. In the CCD, the adiabatic potentials (=the total energy except the atomic kinetic energy, see Eq. (1.4)) are presented as a function of the positions of all the constituent atoms $\{\boldsymbol{R}_j\}$, or a suitable set of the configuration coordinates $\{Q_i\} \equiv Q$, hence multidimensional variables, whose number is basically order of 10^{23}. Figure 3.1(a) illustrates a CCD for a two-atom molecule, which is a simplest example of condensed matter. There are 6 (=3 × 2) freedoms in the atomic positions in this

Electron–Lattice Interactions in Semiconductors
Yuzo Shinozuka

ISBN 978-981-4800-96-9 (Hardcover), 978-1-003-02346-3 (eBook)
www.jennystanford.com

molecule, but it is enough to focus only on the distance Q between two atoms. The lower branch of the adiabatic potential $U_g(Q)$ is for the electronic ground state Ψ_g and the upper $U_{ex}(Q)$ is for the excited state Ψ_{ex} as a function of Q. The minimum position (distance) $\bar{Q}$ of $U_g(Q)$ corresponds to the stable configuration (distance) for Ψ_g, where the effective force $F^g = -\partial U_g(Q)/\partial Q$ between two atoms is zero. On the other hand, in Ψ_{ex} the effective force $F^{ex} = -\partial U_{ex}(Q)/\partial Q$ at $\bar{Q}$ is nonzero because the electronic charge density is different between Ψ_{ex} and Ψ_g. If this molecule in the ground state is excited by photoabsorption, the constituent atoms feel an induced force $F^{ex} \neq 0$ and then move toward to a new stable position for the excited state (sometimes ∞). This is an example of the adiabatic instability, and the root of the electron–lattice interaction in macroscopic materials.

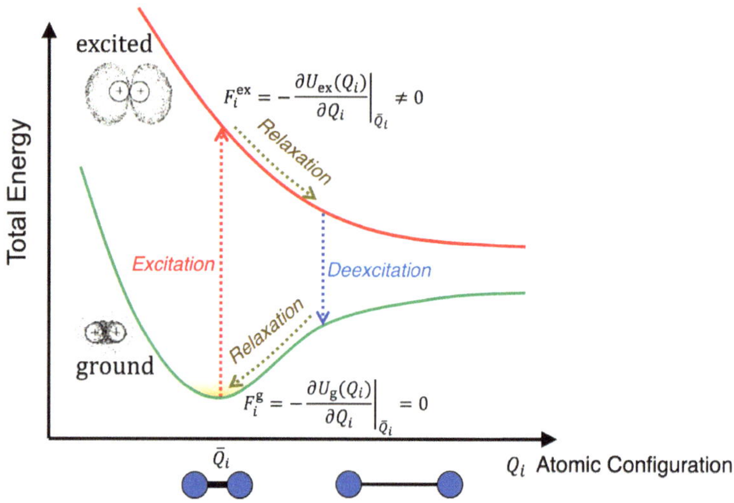

Figure 3.1 Typical configuration coordinate diagram (CCD) for a two-atom molecule. Reproduced with permission from Ref. [1].

3.2 CCD for Two-Level System

Next, let us move to a more complex system: a two-level system in condensed matters. Typical examples are a conduction electron bound at a point defect which has two localized states, Ψ_g (ground) and Ψ_{ex} (excited), interacts with the lattice vibrations $\{q_k\}$ around it (Fig. 3.2a). The following arguments can be applied to many other two-level systems, i.e., either localized or extended one-electron state and also for two different states of many-electron systems. For simplicity, we assume here that both Ψ_g and Ψ_{ex}, do not change along the atomic configurations, at least retain each character (i.e., s-state or p-state).

Based on the discussion in relation with Eq. (2.8), when the electronic system is in the ground state Ψ_g with an electronic energy ε_g, the adiabatic potential can be given by

$$U_g(\{q_k\}) = \varepsilon_g + \frac{1}{2}\sum_k \omega_k^2 q_k^2. \tag{3.1}$$

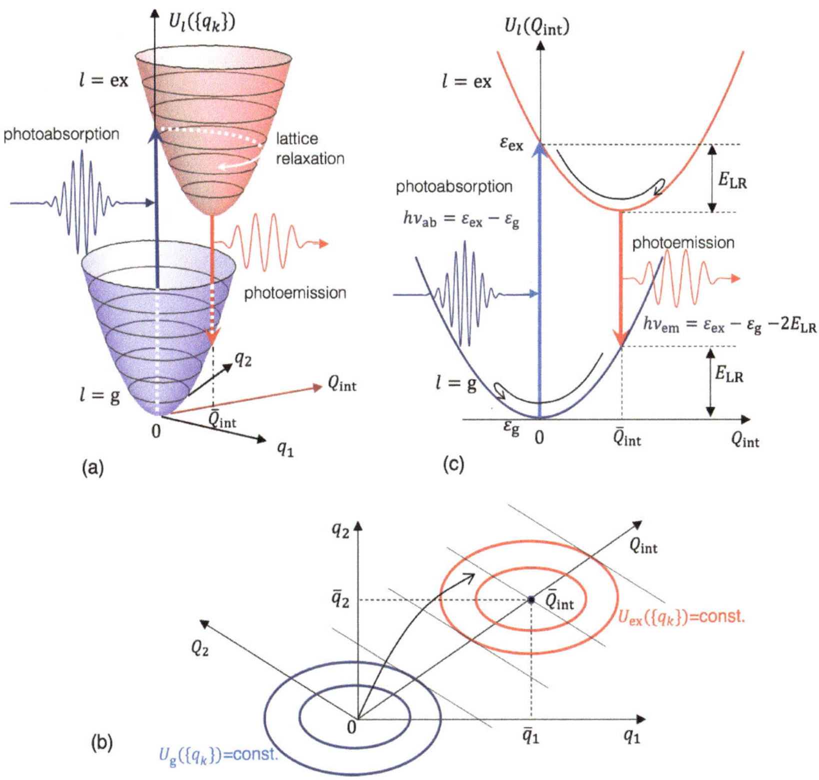

Figure 3.2 The adiabatic potentials for two-level system: (a) illustrated in two normal coordinates q_1 and q_2 in a three-dimensional view, (b) contour map presentation, and (c) illustrated along the interaction mode Q_{int}.

Here q_k is the normal mode of the lattice vibration with an angular frequency ω_k. The subscript k stands all for the related wave vectors, mode index and any localized modes around the defect.[1] The adiabatic potential for the electronic excited state Ψ_{ex} with ε_{ex} can be also given by

$$U_{\text{ex}}(\{q_k\}) = \varepsilon_{\text{ex}} - \sum_k a_k q_k + \frac{1}{2}\sum_k \omega_k^2 q_k^2$$

$$= \varepsilon_{\text{ex}} - \frac{1}{2}\sum_k \omega_k^2 \left(q_k - \frac{a_k}{\omega_k^2}\right)^2 - \frac{1}{2}\sum_k \omega_k^2 \left(\frac{a_k}{\omega_k^2}\right)^2. \quad (3.2)$$

The phonon frequencies may be changed in the excited state, but here we take account of only the difference in the stable positions of the lattice $\{q_k\}$ between two adiabatic potentials $U_{\text{g}}(\{q_k\})$ and $U_{\text{ex}}(\{q_k\})$. The generalized force along the mode q_k at $q_k = 0$ in the excited state is $-\partial U_{\text{ex}}/\partial q_k = a_k$. Accordingly, the minimum position for the mode q_k is shifted to $\bar{q}_k = a_k/\omega_k^2$. Once all the modes have been shifted to the respective minimum as $\{q_k: 0 \rightarrow \bar{q}_k\}$, the total energy is lowered by a quantity $E_{\text{LR}} = (1/2) \Sigma_k a_k^2/\omega_k^2$, which we call the lattice relaxation energy. It should be noted here that the magnitude of the induced force $F_k^{\text{ex}} = -\partial U_{\text{ex}}/\partial q_k$ and the quantity a_k/ω_k^2 are both infinitesimal in each mode, but the total sum over all modes would be large, i.e., typically $E_{\text{LR}} \sim 1$ eV.

Next, we take an orthogonal transformation of the lattice modes from $\{\omega_k q_k\}$ to $\{Q_l\}$ in which one of the modes Q_{int} is chosen as (Fig. 3.2b)

$$Q_{\text{int}} \equiv \bar{Q}_{\text{int}}^{-1} \sum_k a_k q_k, \quad \bar{Q}_{\text{int}} = \bar{Q}_{\text{int}}^{-1} \sum_k a_k \bar{q}_k = \sqrt{\sum_k a_k^2/\omega_k^2}. \quad (3.3)$$

Then the adiabatic potentials in $\{Q_l\}$ space are expressed as

$$U_{\text{g}}(\{Q_l\}) = \varepsilon_{\text{g}} + \frac{1}{2}\sum_l Q_l^2. \quad (3.4)$$

[1]Here we define the origin of q_k at the equilibrium position for the ground state Ψ_{g}, which may be different from those of the reference bulk crystal phonons. After a suitable shift of q_k, within the harmonic approximation nothing will be changed in the following results.

$$U_{ex}(\{Q_l\}) = \varepsilon_{ex} - \bar{Q}_{int}Q_{int} + \frac{1}{2}\sum_l Q_l^2. \tag{3.5}$$

Now the electron–lattice interaction (coupling term) appears only through Q_{int}, which is called the interaction mode hereafter [2, 3]. In $\{Q_l\}$ space, the minimum of $U_{ex}(\{Q_l\})$ is located at $(\bar{Q}_{int}, 0, 0, ...)$ along the Q_{int} direction. Since the other modes $\{Q_{l\neq int}\}$ do not affect electronic energy, the adiabatic potentials are enough to be illustrated only along one-dimensional Q_{int} mode. Here the lattice relaxation energy is given by $(1/2)\bar{Q}_{int}^2$ and the generalized force at the origin is $-\partial U_{ex}/\partial Q_{int}|_0 = -\bar{Q}_{int}$. In other words, the interaction mode is defined so as to connect the minimums for two different electronic states. A typical example of the interaction mode Q_{int} is a breathing mode for two spherically localized electronic states (ground and excited). As can be seen in the definition in Eq. (3.3), the interaction mode Q_{int} consists of many normal modes $\{q_k\}$ with different frequencies, then is (usually) no more a normal mode.

3.3 Lattice Relaxation Dynamics in Two-Level System

Let us discuss the relaxation process, i.e., the time evolution of the lattice within a classical treatment. Assuming that before the photoabsorption ($t < 0$), the electron is in the ground state Ψ_g and the lattice does not move at all from the equilibrium position $\{q_k = 0\}$ ($\{Q_l = 0\}$) [4, 5].

(1) At time $t = 0$ the electronic system is excited by a photon absorption from the ground state Ψ_g to the excited state Ψ_{ex} with an energy $h\nu = \varepsilon_{ex} - \varepsilon_g$. Since the duration of an optical transition is very short ($\sim 10^{-15}$ sec), and much shorter than the typical lattice oscillation periods ($\sim 10^{-13}$ sec), we can assume that the lattice does not move at all during the electronic transition (Franck–Condon principle).

(2) Just after the photoabsorption, the system is in the excited adiabatic potential surface $U_{ex}(\{q_k\})$ and the induced force $F_k^{ex} = -\partial U_{ex}/\partial q_k$ makes the lattice to move from $q_k = 0$ toward a new equilibrium $q_k = \bar{q}_k$, starting with $\dot{q}_k(t = 0) = 0$. Since

all the normal modes q_k are independent with each other, the time dependence is simply given by

$$q_k(t) = \theta(t)\, \bar{q}_k(1 - \cos\omega_k t). \tag{3.6}$$

Here $\theta(t)$ is a step function: $\theta(t) = 0$ for $t < 0$, and $\theta(t) = 1$ for $t \geq 0$. Then the trajectory of the total motion $\{q_k(t)\}$ is a Lissajous in $\{q_k\}$ space, while in $\{Q_l\}$ space, the time dependence of the interaction mode $Q_{\text{int}}(t)$ is given by

$$Q_{\text{int}}(t) = \theta(t)\, \bar{Q}_{\text{int}}^{-1}\, \bar{q}_k \sum_k a_k \bar{q}_k (1 - \cos\omega_k t). \tag{3.7}$$

If the distribution of the phonon frequency $\{\omega_k\}$ has a finite width $\Delta\omega$, the phase in each $\cos\omega_k t$ term will be quickly out of phase, resulting $Q_{\text{int}}(t)$ shows a damping oscillation around $\bar{Q}_{\text{int}}$ (Fig. 3.3). During the dephasing time, $\tau \sim 2\pi/\Delta\omega$, a part of the initial excess electronic energy $\varepsilon_{\text{ex}} - \varepsilon_{\text{g}}$, a quantity

$$E_{\text{LR}} = \frac{1}{2}\bar{Q}_{\text{int}}^2 = \frac{1}{2}\sum_k \left(\frac{a_k}{\omega_k}\right)^2, \tag{3.8}$$

is transformed into the lattice kinetic energy and dissipated over the crystal. This is why the quantity (3.8) is called the lattice relaxation energy. Concurrently the same amount of energy has been stored around the defect as a static elastic energy U_L (=E_{LR}). It should be noted that in the normal mode $\{q_k\}$ space, each $q_k(t)$ simply oscillates independently and there is no energy dissipation at all. The dimensionless quantity $S = E_{\text{LR}}/\hbar\bar{\omega}_k$, where $\bar{\omega}_k$ is the average frequency, is called the Huang–Rhys factor [6], representing the average number of generated phonons (in quantum treatment) during the relaxation. It can be shown that for $S >> 1$ above classical treatment of the lattice motion can be applicable.

If there is a local lattice vibration mode (=normal mode) at the defect and a condition $a_{\text{loc}}\bar{q}_{\text{loc}} \sim 1$ is satisfied, the transient lattice vibration lasts even after $t > \tau$,

$$Q_{\text{int}}(t) \approx \frac{1}{\bar{Q}_{\text{int}}} a_{\text{loc}}\bar{q}_{\text{loc}}(1 - \cos\omega_{\text{loc}} t). \tag{3.9}$$

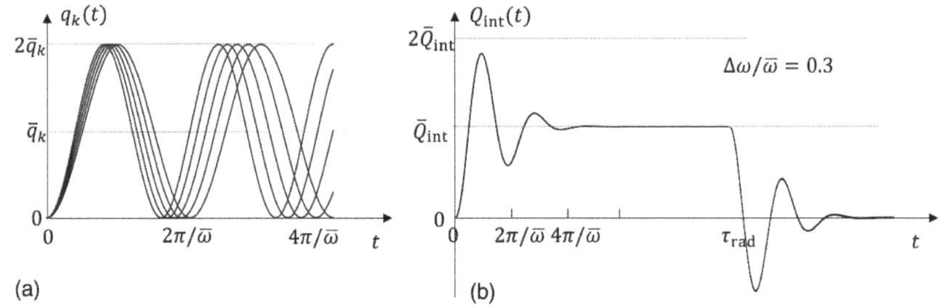

Figure 3.3 (a) Dephasing among different normal modes $q_k(t)$. (b) Time evolution of the interaction modes $Q_{int}(t) = \bar{Q}_{int}^{-1}\Sigma_k a_k q_k(t)$: assuming a Gaussian distribution of $\{\omega_k\}$ with $\Delta\omega/\omega = 0.3$.

Except for this local mode q_{loc}, the amount of lattice relaxation energy $E'_{LR} = (1/2)\Sigma_{k\neq loc}(a_k/\omega_k)^2$ is dissipated over the crystal during $0 < t < \tau$. The rest energy, $E_{LR} - E'_{LR}$, which is once stored in the local vibration mode, is gradually dissipated through unharmonic terms which has not been considered in Eq. (3.2).

Let us next discuss a photoemission process from Ψ_{ex} to Ψ_g which occurs at time $\tau_{rad}(\gg\tau)$. The radiative electronic lifetime τ_{rad} is usually the order of $\sim 10^{-9}$ much longer than the lattice relaxation time τ ($\sim 10^{-12}$ sec). Then before the photoemission, the transient lattice vibration generated by the photoabsorption has been completely damped out. According to the Franck–Condon principle as same as above, a photon is emitted with an energy $\varepsilon_{ex} - \varepsilon_g - (E_{LR} + U_L)$. The energy difference between the absorption and emission is $(E_{LR} + U_L)$, called the Stokes shift. Just after the photoemission, the lattice begins to move in the $U_g(\{q_k\})$ space from $q_k = \bar{q}_k$ toward a new equilibrium $q_k = 0$. The time evolution of the lattice is shown to be

$$Q_{int}(t) = \theta(t)\, \bar{Q}_{int}^{-1}\, \bar{q}_k \sum_k a_k \bar{q}_k (1 - \cos\omega_k t)$$

$$+\theta(t - \tau_{rad})\, \bar{Q}_{int}^{-1}\, \bar{q}_k \sum_k a_k \bar{q}_k \cos\omega_k (t - \tau_{rad}). \quad (3.10)$$

The first term has been generated by the photoabsorption at $t = 0$, and hence has almost damped out for $t \gg \tau$. The second term generated by the photoemission at $t = \tau_{rad}$ is the main term for $t > \tau_{rad}$, which shows a damping oscillation around $Q_{int} = 0$

(Fig. 3.3). During the damping oscillation, the restored elastic energy U_L (=E_{LR}) is dissipated as a kinetic energy of the atomic motion in a period τ (=$2\pi/\Delta\omega$). After all, among the initial excess electronic energy $\varepsilon_{ex} - \varepsilon_g$, the amount of E_{LR}, is dissipated during the relaxation triggered by the first transition (absorption) and the amount of U_L (=E_{LR}) is dissipated during the relaxation after the second transition (emission).

3.4 Photoabsorption and Photoemission (Classical and Statistical Treatment)

So far, we have discussed the lattice dynamics induced by photoabsorption and photoemission at absolute zero temperature $T = 0$. Let us next discuss these optical processes at finite temperature $T \neq 0$ K with a statistical and classical treatment of the lattice, which is applicable for a situation $\varepsilon_{ex} - \varepsilon_g \gg k_B T$ [3].

3.4.1 Photoabsorption

Starting from a situation where the electronic system is in the ground state Ψ_g, while the lattice vibrations $\{q_k\}$ are excited thermally (randomly) around $\{q_k = 0\}$: $\varepsilon_{ex} - \varepsilon_g \gg k_B T$ (Fig. 3.4). Since the optical transition takes place (~fsec) much shorter than the typical time constant of the lattice motion ($\sim 2\pi/\omega_k$), it would be acceptable to assume that the lattice $\{q_k\}$ does not move during the electronic transition: the Franck–Condon principle. The value of $\{q_k\}$ before the photoabsorption distributes randomly and could be given by the Boltzmann distribution as

$$f_g(\{q_k\}) = A \exp\left[-\frac{U_g(\{q_k\})}{k_B T}\right] = A' \prod_k \exp\left[-\frac{\omega_k^2 q_k^2}{2\, k_B T}\right], \quad (3.11)$$

where the normalization condition is

$$\int_{-\infty}^{\infty} \cdots \int_{-\infty}^{\infty} f_g(\{q_k\})\, dq_{k1} \cdots dq_{kN} = 1. \quad (3.12a)$$

$$f_g(\{q_k\}) = \prod_k \frac{1}{\omega_k \sqrt{2\pi k_B T}} \exp\left[-\frac{\omega_k^2 q_k^2}{2\, k_B T}\right]. \quad (3.12b)$$

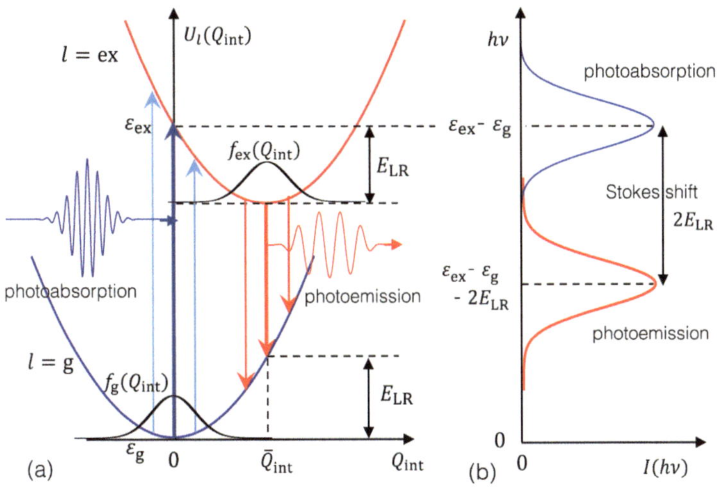

Figure 3.4 (a) adiabatic potentials of a two-level system. (b) photoabsorption and photoemission spectrum.

Using the Fermi's golden rule, the optical transition probability for a single system absorbing a photon $h\nu$ is given by

$$I_{ab}^{single}(h\nu) = \frac{2\pi}{\hbar}\left|\langle\psi_{ex}|H_{int}|\psi_g\rangle\right|^2 \times \delta\big(U_{ex}(\{q_k\})-U_g(\{q_k\}) - h\nu\big). \tag{3.13}$$

Here, $|\psi_g\rangle$ denotes the electronic ground state and $|\psi_{ex}\rangle$ denotes the electronic excited state. $\delta(x)$ is the Dirac δ-function representing the energy conservation, and

$$M_{ex,g} \equiv \langle\psi_{ex}|H_{int}|\psi_g\rangle, \tag{3.14}$$

is the matrix element of the interaction between electron and the radiation field.

Optical measurements are usually done on a sample that contains with many similar subsystems, then multiplying the distribution $f_g(\{q_k\})$ with Eq. (3.14) and integrating over all the mode $\{q_k\}$, the photoabsorption spectrum is given by

$$I_{ab}(h\nu) = \frac{2\pi}{\hbar}\int_{-\infty}^{\infty}\cdots\int_{-\infty}^{\infty} f_g(\{q_k\})\left|M_{ex,g}\right|^2 \times \delta\big(U_{ex}(\{q_k\})-U_g(\{q_k\}) - h\nu\big)dq_{k1}\cdots dq_{kN}, \tag{3.15}$$

Substituting Eqs. (3.1) and (3.2) into (3.15) we obtain

$$I_{\text{ab}}(h\nu) = \frac{2\pi}{\hbar}\left|M_{\text{ex,g}}\right|^2 \left(\prod_k \frac{1}{\omega_k\sqrt{2\pi k_B T}}\right)\int_{-\infty}^{\infty}\cdots\int_{-\infty}^{\infty}\exp\left[-\frac{\sum_k \omega_k^2 q_k^2}{2\,k_B T}\right]$$
$$\times\,\delta\left(\varepsilon_{\text{ex}}-\varepsilon_{\text{g}}-\textstyle\sum_k a_k q_k - h\nu\right)dq_{k1}\cdots dq_{kN}. \qquad (3.16)$$

Next, move from $\{\omega_k q_k\}$-space to $\{Q_l\}$-space with a Jacobian $\prod_k \omega_k$,

$$I_{\text{ab}}(h\nu) = \frac{2\pi\left|M_{\text{ex,g}}\right|^2}{\hbar(\sqrt{2\pi k_B T})^N}\int_{-\infty}^{\infty}\exp\left[-\frac{Q_{\text{int}}^2}{2k_B T}\right]\delta\left(\varepsilon_{\text{ex}}-\varepsilon_{\text{g}}-\bar{Q}_{\text{int}}Q_{\text{int}}-h\nu\right)dQ_{\text{int}}$$
$$\times\int_{-\infty}^{\infty}\exp\left[-\frac{Q_2^2}{2k_B T}\right]dQ_2\cdots\int_{-\infty}^{\infty}\exp\left[-\frac{Q_N^2}{2\,k_B T}\right]dQ_N.$$
$$= \frac{2\pi\left|M_{\text{ex,g}}\right|^2}{\hbar}\frac{1}{\bar{Q}_{\text{int}}\sqrt{2\pi k_B T}}\exp\left[-\frac{\left(h\nu-(\varepsilon_{\text{ex}}-\varepsilon_{\text{g}})\right)^2}{2\bar{Q}_{\text{int}}^2 k_B T}\right]. \qquad (3.17)$$

The absorption spectrum is thus turned out to be a Gaussian centered at ε_{ex} - ε_{g}. The spectrum width is given by $\bar{Q}_{\text{int}}\sqrt{2k_B T}$, which is called a homogeneous broadening, originated from the thermal lattice-fluctuation, in contrast to inhomogeneous broadening usually originated from the statistical fluctuation of the central energy level, $\varepsilon_{\text{ex}} - \varepsilon_{\text{g}}$.

In analysis of the temperature dependence of the optical spectra obtained by experiments, it is often to introduce the effective temperature T_{eff} defined by

$$k_B T_{\text{eff}} \equiv \frac{\hbar\omega}{2}\coth\left(\frac{\hbar\omega}{2k_B T}\right). \qquad (3.18)$$

to fit the spectra with the Gaussian Eq. (3.17). At high temperature $\hbar\omega \ll k_B T$, it reduces $T_{\text{eff}} \sim T$ where the classical treatment is valid. At low temperature $\hbar\omega \gg k_B T$, on the other hand, the broadening of the spectrum is considered to be originated from the zeropoint vibration: $k_B T_{\text{eff}} = \hbar\omega/2$.

3.4.2 Photoemission (Classical Treatment)

After the transient lattice relaxation induced by a photoabsorption has been completed, the lattice motions in the electronic excited state would be thermally activated around the minimum $\{\bar{q}_k\}$. The distribution is given by

$$f_{\mathrm{ex}}(\{q_k\}) = \prod_k \frac{1}{\omega_k\sqrt{2\pi k_{\mathrm{B}}T}} \exp\left[-\frac{\omega_k^2(q_k-\bar{q}_k)^2}{2\,k_{\mathrm{B}}T}\right]. \tag{3.19}$$

With a similar procedure as in the photoabsorption, the photoemission spectrum is shown to be

$$I_{\mathrm{em}}(h\nu) = \int_{-\infty}^{\infty}\cdots\int_{-\infty}^{\infty} f_{\mathrm{ex}}(\{q_k\})\left|\langle\psi_{\mathrm{g}}|H_{\mathrm{int}}|\psi_{\mathrm{ex}}\rangle\right|^2$$

$$\times\,\delta\big(U_{\mathrm{ex}}(\{q_k\})-U_{\mathrm{g}}(\{q_k\})-h\nu\big)dq_{k1}\cdots dq_{kN}$$

$$= \frac{2\pi\left|M_{\mathrm{g,ex}}\right|^2}{\hbar^2}\frac{1}{\bar{Q}_1\sqrt{2\pi k_{\mathrm{B}}T}}\exp\left[-\frac{\{h\nu-(\varepsilon_{\mathrm{ex}}-\varepsilon_{\mathrm{g}}-\bar{Q}_{\mathrm{int}}^2)\}^2}{2\bar{Q}_1^2 k_{\mathrm{B}}T}\right]. \tag{3.20}$$

The emission spectrum is a Gaussian centered at $\varepsilon_{\mathrm{ex}} - \varepsilon_{\mathrm{g}} - \bar{Q}_{\mathrm{int}}^2$ with a homogeneous width $\bar{Q}_{\mathrm{int}}\sqrt{2k_{\mathrm{B}}T}$. The energy difference in the peaks of the absorption and the emission is given by $\bar{Q}_{\mathrm{int}}^2 = 2E_{\mathrm{LR}}$, which is called the Stokes shift. As have discussed, a half of the Stokes shift is due to the lattice relaxation in the excited state and the other half is stored as a lattice elastic energy around the system. Thus, the electronic energy $2E_{\mathrm{LR}}$ is dissipated over the lattice by two relaxation steps.

It is worthy to note that the Einstein relation holds for the absorption and emission spectra in this system [3]:

$$I_{\mathrm{em}}(h\nu) = \frac{\left|M_{\mathrm{g,ex}}\right|^2}{\left|M_{\mathrm{ex,g}}\right|^2}\exp\left(-\frac{h\nu}{k_{\mathrm{B}}T}\right)I_{\mathrm{ab}}(h\nu). \tag{3.21}$$

This useful, and somewhat surprising, relation holds for any systems where the absorption and emission processes start from each thermal equilibrium and no correlation between them.

If we treated the radiation field classical, the matrix elements appearing in Eq. (3.21) would be the same and then cancel each other. On the other hand, if we treat the radiation field as quantum variables, each matrix element is related to so-called Einstein's A and B coefficients, then the ratio is shown to be $4h\nu^3/c^3$. A general proof is given in Appendix II with Einstein's A and B coefficients.

So far, we have assumed that the photoabsorption and photoemission processes are, respectively, a first-order optical process with respect to an electronic transition, where the two processes can be treated separately. However, if the two processes occur within a very short duration (order of fsec (10^{-15} sec)), the quantum coherence should be crucial and treated quantum mechanically. This second-order optical process is discussed in detail in Ref. [3].

3.5 Photoabsorption and Photoemission at a Localized Electron System (Quantum Treatment)

3.5.1 Coupled with a Single Lattice Mode

In this section, we study a simple case where a localized electronic system coupled with a single lattice vibration mode, which will be treated in quantum mechanics (Fig. 3.5). A typical example is an electron localized at a vacancy or impurity in a semiconductor, coupled with a localized vibration mode. Here we assume that the electronic system has two states, the ground $|\psi_g\rangle$ and excited $|\psi_{ex}\rangle$ with eigenenergy ε_g and ε_{ex}, respectively. The coupled lattice coordinate is q and its conjugate momentum p with an angular frequency ω with a relevant mass M. The equilibrium position of the lattice is $q = 0$ for $|\psi_g\rangle$ and $q = \bar{q}$ for $|\psi_{ex}\rangle$.

The Hamiltonian we consider is

$$H = |\psi_g\rangle\left(\frac{p^2}{2M} + \frac{1}{2}M\omega^2 q^2\right)\langle\psi_g|$$

$$+|\psi_{ex}\rangle\left(\frac{p^2}{2M} + \frac{1}{2}M\omega^2(q-\bar{q})^2 - \frac{1}{2}M\omega^2\bar{q}^2\right)\langle\psi_{ex}|. \quad (3.22)$$

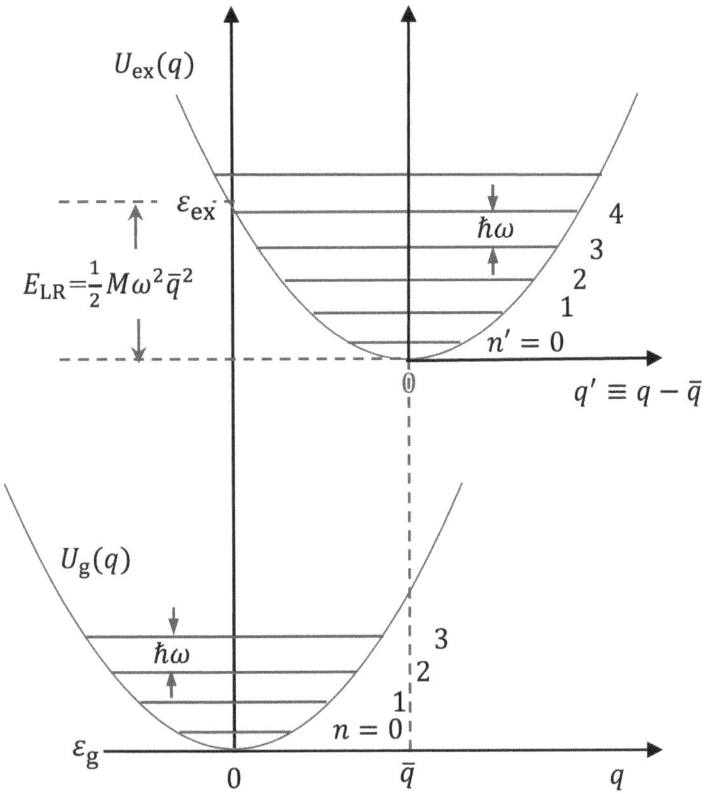

Figure 3.5 A two-level electronic system coupled with a single normal lattice mode q.

Electron ground state

The eigenstates are direct products of the electron part and the phonon part:

$$|\psi_g\rangle \times |\chi^L_{g,n}\rangle, \quad E_{g,n} = \varepsilon_g + \left(n + \frac{1}{2}\right)\hbar\omega, \quad n = 0, 1, 2, \dots \tag{3.23}$$

$$\chi^L_{g,n}(q) = N_n H_n(\alpha q)\exp\left[-\frac{1}{2}\alpha^2 q^2\right], \text{ with } \alpha \equiv \sqrt{\frac{M\omega}{\hbar}}. \tag{3.24}$$

Here $H_n(z)$ is the n-th Hermite polynomial, defined as

$$H_n(z) = (-1)^n e^{z^2} \frac{d^n}{dz^n} e^{-z^2}. \tag{3.25}$$

Electron excited state

The equilibrium position of the lattice is shifted to $\bar{q}$, then $q' \equiv q - \bar{q}$ will be a suitable variable for the phonon state in the electron excited state. The eigenstate and energy are

$$|\psi_{\rm ex}\rangle \times |\chi^{\rm L}_{{\rm ex},n'}\rangle,\ E_{{\rm ex},n'} = \varepsilon_{\rm ex} + \left(n' + \frac{1}{2}\right)\hbar\omega - E_{\rm LR},\ \ n' = 0, 1, 2, \ldots \tag{3.26}$$

$$\chi^{\rm L}_{{\rm ex},n'}(q') = N_{n'}H_{n'}(\alpha q')\exp\left[-\frac{1}{2}\alpha^2 q'^2\right]. \tag{3.27}$$

Here the lattice relaxation energy is $E_{\rm LR} = (1/2)M\omega^2\bar{q}^2$.

Photoabsorption

Before the photoabsorption, we assume that thermal equilibrium has been achieved among phonon states $|\chi^{\rm L}_{{\rm g},n}\rangle (n = 0, 1, 2, \ldots)$ in the electron ground state. The photoabsorption process consists of various transitions between phonon states $(g, n) \to ({\rm ex}, n')$

$$I_{\rm abs}(h\nu) = \frac{2\pi}{\hbar^2}\sum_n\sum_{n'} W_n \left|\langle\psi_{\rm ex} \times \chi^{\rm L}_{{\rm ex},n'}|H_{\rm int}|\psi_{\rm g} \times \chi^{\rm L}_{{\rm g},n}\rangle\right|^2 \delta\left(E_{{\rm ex},n'} - E_{{\rm g},n} - h\nu\right) \tag{3.28}$$

with the Boltzmann factor referring to the electron-ground state.

$$W_n = \exp\left[-E_{{\rm g},n}/k_BT\right] / \sum_n \exp\left[-E_{{\rm g},n}/k_BT\right] \tag{3.29}$$

Since the perturbation $H_{\rm int}$ by radiation fields usually contains only electronic variable,

$$\left|\langle\psi_{\rm ex} \times \chi^{\rm L}_{{\rm ex},n'}|H_{\rm int}|\psi_{\rm g} \times \chi^{\rm L}_{{\rm g},n}\rangle\right|^2 = \left|\langle\psi_{\rm ex}|H_{\rm int}|\psi_{\rm g}\rangle\right|^2 \left|\langle\chi^{\rm L}_{{\rm ex},n'}|\chi^{\rm L}_{{\rm g},n}\rangle\right|^2$$

$$\equiv \left|M_{\rm ex,g}\right|^2 \times \left|\langle\chi^{\rm L}_{{\rm ex},n'}|\chi^{\rm L}_{{\rm g},n}\rangle\right|^2. \tag{3.30}$$

Then

$$I_{\rm abs}(h\nu) = \frac{2\pi}{\hbar^2}\left|M_{\rm ex,g}\right|^2 \sum_n\sum_{n'} W_n \left|\langle\chi^{\rm L}_{{\rm ex},n'}|\chi^{\rm L}_{{\rm g},n}\rangle\right|^2 \delta\left(E_{{\rm ex},n'} - E_{{\rm g},n} - h\nu\right). \tag{3.31}$$

Photoemission

After thermal equilibrium among phonon states $|\chi^{\mathrm{L}}_{\mathrm{ex},n'}\rangle$ (n' = 0, 1, 2, ...) in the electronic excited state has been completed, the photoemission process consists of various transitions $(\mathrm{ex}, n') \to (g, n)$

$$I_{\mathrm{em}}(h\nu) = \frac{2\pi}{\hbar^2} \left|M_{\mathrm{g,ex}}\right|^2 \sum_{n'} \sum_{n} W_{n'} \left|\langle \chi^{\mathrm{L}}_{\mathrm{g},n} | \chi^{\mathrm{L}}_{\mathrm{ex},n'} \rangle\right|^2 \delta\left(E_{\mathrm{ex},n'} - E_{\mathrm{g},n} - h\nu\right). \tag{3.32}$$

with the Boltzmann factor referring to the electron-excited state.

$$W_{n'} = \exp\left[-E_{\mathrm{ex},n'}/k_B T\right] / \sum_{n'} \exp\left[-E_{\mathrm{ex},n'}/k_{\mathrm{B}} T\right]. \tag{3.33}$$

***T* = 0 K**

At absolute zero temperature, the initial state is restricted to the zero phonon state: (g, n = 0) for absorption and (ex, n' = 0) for emission. After some calculations on the matrix element $\langle \chi^{\mathrm{L}}_{\mathrm{ex},n'} | \chi^{\mathrm{L}}_{\mathrm{g},n} \rangle$ (see Appendix III), we obtain the following formulas:

$$I_{\mathrm{abs}}(h\nu) \propto \sum_{n'=0} \frac{S^{n'}}{n'!} \exp[-S]\, \delta\left(\varepsilon_{\mathrm{ex}} - \varepsilon_{\mathrm{g}} - S\hbar\omega + n'\hbar\omega - h\nu\right), \tag{3.34}$$

$$I_{\mathrm{em}}(h\nu) \propto \sum_{n=0} \frac{S^{n}}{n!} \exp[-S]\, \delta\left(\varepsilon_{\mathrm{ex}} - \varepsilon_{\mathrm{g}} - S\hbar\omega - n\hbar\omega - h\nu\right), \tag{3.35}$$

with the Huang–Rhys factor [6]

$$S \equiv \frac{E_{\mathrm{LR}}}{\hbar\omega}, \quad E_{\mathrm{LR}} \equiv \frac{1}{2} M\omega^2 \bar{q}^2, \tag{3.36}$$

which denotes the strength of the electron–phonon coupling. Both spectra consist of a set of line spectrum of a Poisson distribution with equal distance $\hbar\omega$. The transition $n = 0 \leftrightarrow n' = 0$ is called the zero-phonon line whose strength is proportional to $\exp[-S]$. The maximum strength in the spectrum is approximately located at $n = S$ ($n' = S$), this corresponds to the vertical transition in the classical model (Section 3.4) after the Franck–Condon principle. Figure 3.6 shows typical examples of spectra. It can be shown that for large $S \gg 1$ the absorption and emission spectra are reshaped into a Gaussian given by Eqs. (3.17) and (3.19), respectively.

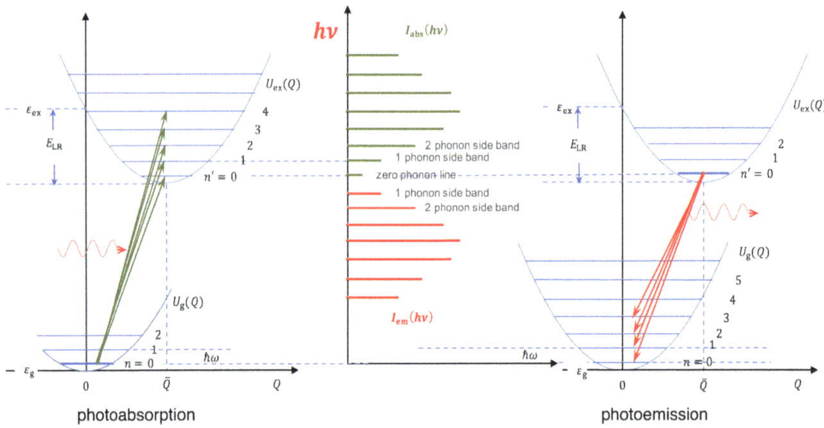

Figure 3.6 CC diagram for a two-level system indicating the photo-absorption and photoemission processes. Both photospectra at $T = 0$ K show the Poisson distribution starting with the zero phonon line.

For finite temperature, $T \neq 0$ K, we must perform the summation over various phonon numbers: Further analysis is given in Ref. [3].

3.5.2 Coupled with Many Lattice Modes

So far, we have considered those systems where an electronic system couples with only a single lattice mode. In order to discuss the coupling with many lattice modes, an extension would be straightforward using the following Hamiltonian form as

$$H = |\Psi_g\rangle \left[\varepsilon_g + \frac{1}{2}\sum_k (p_k^2 + \omega_k^2 q_k^2)\right] \langle\Psi_g|$$

$$+|\Psi_{ex}\rangle \left[\varepsilon_{ex} - \sum_k a_k q_k + \frac{1}{2}\sum_k (p_k^2 + \omega_k^2 q_k^2)\right] \langle\Psi_{ex}|. \quad (3.37)$$

The eigenstates are a direct product of the electron part and the phonon part.

Electron ground state: $|\Psi_g\rangle|\chi^{L}_{g,\{n_k\}}\rangle = |\Psi_g\rangle \prod_k \frac{1}{\sqrt{n_k}}(b_k^{+})^{n_k}|0\rangle,$

(3.38a)

$$E_{g,\{n_k\}} = \varepsilon_g + \sum_k \hbar\omega_k\left(n_k + \frac{1}{2}\right). \qquad n_k = 0, 1, 2, \ldots \tag{3.38b}$$

Electron excited state: $|\Psi_{ex}\rangle|\chi^{L}_{g,\{n_k\}}\rangle = |\Psi_{ex}\rangle \prod_{k'} \frac{1}{\sqrt{n_{k'}}}(b'^{+}_{k'})^{n_{k'}}|0'\rangle,$ (3.39a)

$$E_{ex,\{n_{k'}\}} = \varepsilon_{ex} + \sum_{k'} \hbar\omega_{k'}(n_{k'} + \frac{1}{2}) - E_{LR}, \quad n_{k'} = 0, 1, 2, \ldots \tag{3.39b}$$

Here $|0\rangle$ and $|0'\rangle$ are the phonon vacuum in the electron ground state and excited state, respectively. The lattice relaxation energy is given by

$$E_{LR} = \frac{1}{2}\sum_k \omega_k^2 \bar{q}_k^2, \qquad \bar{q}_k = a_k/\omega_k^2. \tag{3.40}$$

The spectrum for a photon $h\nu$ absorption consists of many transitions between various phonon states (g, $\{n_k\}$) → (ex, $\{n_{k'}\}$).

$$I_{abs}(h\nu) = \frac{2\pi}{\hbar^2}|M_{ex,g}|^2 \sum_k \sum_{n_k} \sum_{k'} \sum_{n_{k'}} W_{g,\{n_k\}} \left|\langle\chi^{L}_{ex,\{n_{k'}\}}|\chi^{L}_{g,\{n_k\}}\rangle\right|^2$$
$$\times\, \delta\left(E_{ex,\{n_{k'}\}} - E_{g,\{n_k\}} - h\nu\right). \tag{3.41}$$

with a Boltzmann factor

$$W_{g,\{n_k\}} = \exp\left[-E_{g,\{n_k\}}/k_B T\right] / \sum_k \sum_{n_k} \exp\left[-E_{g,\{n_k\}}/k_B T\right]. \tag{3.42}$$

For photoemission after thermal equilibrium in the electronic excited state, transitions occur between various phonon states (ex, $\{n_{k'}\}$) → (g, $\{n_k\}$)

$$I_{em}(h\nu) = \frac{2\pi}{\hbar^2}|M_{g,ex}|^2 \sum_k \sum_{n_k} \sum_{k'} \sum_{n_{k'}} W_{ex,\{n_{k'}\}} \left|\langle\chi^{L}_{g,\{n_k\}}|\chi^{L}_{ex,\{n_{k'}\}}\rangle\right|^2$$
$$\times\, \delta\left(E_{ex,\{n_{k'}\}} - E_{g,\{n_k\}} - h\nu\right) \tag{3.43}$$

with a Boltzmann factor

$$W_{\text{ex},\{n_{k'}\}} = \exp\left[-E_{\text{ex},\{n_{k'}\}}/k_{\text{B}}T\right] / \sum_{n'} \exp\left[-E_{\text{ex},\{n_{k'}\}}/k_{\text{B}}T\right]. \quad (3.44)$$

Further detailed analysis is found at Refs. [3, 7], which can be summarized as follows:

Let define the respective Huan–Rhys factor for the k-th phonon mode

$$s_k \equiv \frac{\omega_k^2 \bar{q}_k^2}{2\hbar\omega_k}, \quad S = \sum_k s_k \hbar\omega_k \quad \text{and} \quad g = \sum_k s_k \coth\left(\frac{\hbar\omega_k}{2k_{\text{B}}T}\right)$$

Also define

$$g(\omega) = \sum_k s_k\{(\bar{n}_k + 1)\delta(\omega - \omega_k) + \bar{n}_k\delta(\omega + \omega_k)\} \quad (3.45)$$

with the average phonon number at temperature T

$$\bar{n}_k = \langle n_k \rangle = \frac{1}{\exp\left(\hbar\omega_k/k_{\text{B}}T\right) - 1}. \quad (3.46)$$

The first term of (3.45) corresponds to phonon emission process and the second term to the phonon absorption process. It is noted that $g = \int g(\omega)d\omega$.

The absorption spectrum is shown to have a form

$$I_{\text{ab}}(h\nu) = \mathrm{e}^{-g} \sum_{n=1}^{\infty} \int d\omega_1 \cdots \int d\omega_n \delta(h\nu_{\text{zero}} + \hbar\omega_1 + \cdots \hbar\omega_n - h\nu)$$

$$\times \frac{g(\omega_1) \cdots g(\omega_n)}{n!}. \quad (3.47)$$

The spectrum has following properties:

(a) The absorption spectrum at the lowest energy $E_{\text{ex},\{0\}} - E_{\text{g}\{0\}} \equiv h\nu_{\text{zero}}$ is a sharp line called the zero-phonon line whose strength is given by e^{-g}, which is called the Debye–Waller factor.

(b) The spectrum consists of phonon side bands located at both sides of the zero-phonon line: The upper side corresponds to phonon emission and the lower side to phonon absorption. The strength of the n-th phonon side band is $e^{-g}g^n/n!$.

(c) In weak coupling case ($g < 1$), the strength of the side bands is weak and quickly decreases for large n.

(d) In strong coupling case ($g \gg 1$), the strength of the zero-phonon side band is very weak. The contribution of each phonon side band is combined to form a broad spectrum

$$I_{\text{abs}}(h\nu) = \frac{1}{\sqrt{2\pi}D}\exp\left[-\frac{\{h\nu - (\varepsilon_{\text{ex}} - \varepsilon_{\text{g}} - E_{\text{LR}})\}^2}{D^2}\right], \tag{3.48}$$

with

$$D \equiv \left[\sum_k (\hbar\omega_k)^2 s_k \coth\left(\frac{\hbar\omega_k}{2k_{\text{B}}T}\right)\right]^{\frac{1}{2}}. \tag{3.49}$$

At high temperature, D becomes $2Sk_{\text{B}}T$; then the spectrum (3.48) tends to the Gaussian broad spectrum given by Eq. (3.17) in the classical situation.

3.6 Nonradiative Transition

When the radiative lifetime τ_{rad} is short and comparable to the relaxation time, τ, a hot luminescence may occur during the transient lattice vibration with emitting a photon $h\nu = U_{\text{ex}}(Q_{\text{int}}(t)) - U_{\text{g}}(Q_{\text{int}}(t))$ along the Franck–Condon principle (Fig. 3.7(a)).

In the excited state, the lattice $Q_{\text{int}}(t)$ vibrates around $\bar{Q}_{\text{int}}$ and can have chances to reach or cross the crossing point $Q^{\text{c}}_{\text{int}}$ where $U_{\text{ex}}(Q^{\text{c}}_{\text{int}}) = U_{\text{g}}(Q^{\text{c}}_{\text{int}})$, i.e., the Franck–Condon energy is zero [8]. In other words, a nonradiative transition from Ψ_{ex} to Ψ_{g} is possible at $Q^{\text{c}}_{\text{int}}$.

There are two alternative approaches to discuss the nonradiative transition between two electronic states, $|\psi_1\rangle$ and $|\psi_2\rangle$, connected with the electronic transfer energy b_{12}, and both couple with the classical coordinate Q (Fig. 3.8).

(a) The system is treated in the adiabatic approximation. Two original adiabatic potentials $U_1(Q)$ and $U_2(Q)$ repel each other at the crossing point due to the electronic transfer energy b_{12} to reconstruct new adiabatic potentials: the upper branch $U_{up}(Q)$ and the lower branch $U_{low}(Q)$ (Fig. 3.8a).

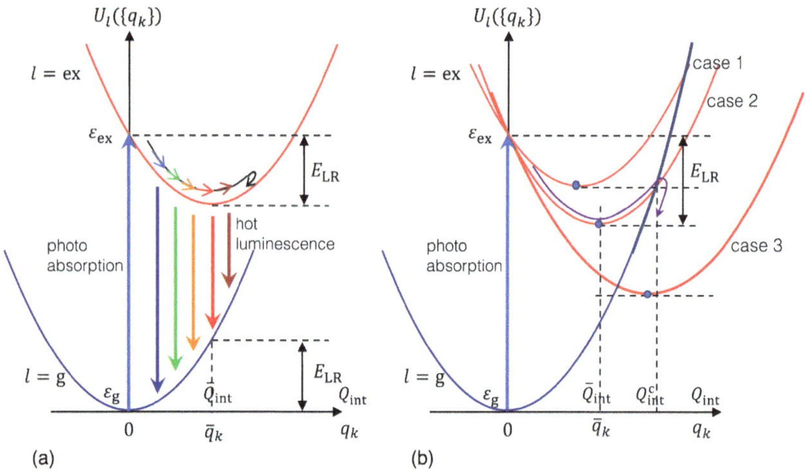

Figure 3.7 (a) Hot luminescence occurs during the lattice relaxation. (b) Nonradiative transition is possible in case 2 and 3. See Appendix VIII.

Consider the initial state is $|\psi_1\rangle$.

If the velocity of $Q(t)$ is slow, the system moves along $U_{up}(Q)$ and the final electronic state is $|\psi_2\rangle$; this is called adiabatic process.

If the velocity of $Q(t)$ is fast, a transition is possible from $U_{up}(Q)$ to $U_g(Q)$ around Q^c_{int}, whose probability can be estimated by the Fermi's golden rule taking the nonadiabatic term H' (Eq. (A1.10) in Appendix I) as a transition matrix element.

(b) The electronic transition matrix b_{12} is neglected at first. Then $U_1(Q)$ and $U_2(Q)$ simply cross each other at the crossing point. Treating the electron transfer b_{12} as a perturbation, we can calculate the transition rate from $U_1(Q)$ to $U_2(Q)$, for example, by the Fermi's golden rule. Consider the initial state is $|\psi_1\rangle$.

If the rate is very small, as the system moves along $U_1(Q)$, nothing happens and the final state is $|\psi_1\rangle$; this is called diabatic process.

If the rate is finite, a transition is possible from $U_1(Q)$ to $U_2(Q)$ around Q^c_{int} resulting in the final state $|\psi_2\rangle$.

Which is appropriate depends on the parameters. The Landau–Zener–Stückelberg model would be helpful for quantitative consideration (Appendix. IV).

Once a nonradiative transition dynamically occurs at the crossing point Q^c_{int} without the activation energy, the initial excess electronic energy $\varepsilon_{ex} - \varepsilon_g$ is dissipated in a very short period τ. The probability depends on the relative positions of $\bar{Q}_{int}$, Q^c_{int}, and τ (Fig. 3.7a, case 2). On the other hand, after the lattice relaxation in the excited state has been completed ($t \sim \tau$), the lattice can be thermally excited to vibrate around $Q_{int}(t) = \bar{Q}_{int}$. In this case, a nonradiative transition needs an excitation energy $(\bar{Q}_{int} - Q^c_{int})^2/2$; hence, the probability shows an activation type.

When the lattice relaxation energy is so large that $\varepsilon_{ex} - \varepsilon_g - \bar{Q}^2_{int} < 0$, radiative transitions from the relaxed electronic excited state ($\bar{Q}_{int} = \{\bar{q}_k\}$) are prohibited, then the relaxed excited state has a long lifetime (Fig. 3.7(b), case 3). These metastable states with a long lifetime will be discussed in Chapter 9 in relation with photo-structural change.

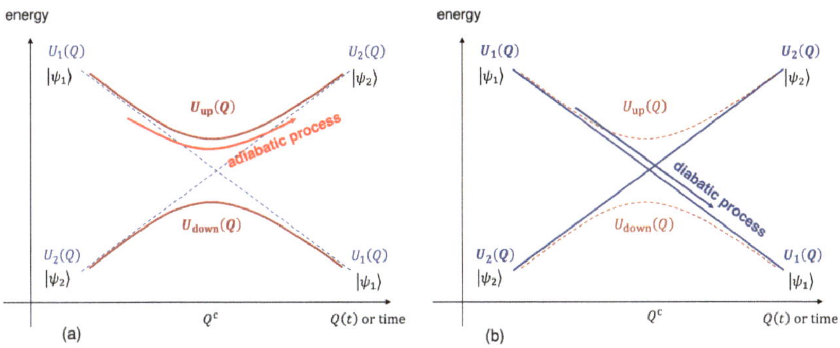

Figure 3.8 Two alternative approaches to nonadiabatic transition between $|\psi_1\rangle$ and $|\psi_2\rangle$. Here $U_{1,2}(Q)$ are illustrated by omitting remaining terms such as $V_{ion}(Q)$.

3.7 Physics of Relaxation

We have learned in Section 3.3 that the lattice relaxation process is related to the dephasing in the relevant many phonon modes.

If the time reversal is possible, anti-relaxation processes could be realized. Is it so?

Let us consider the transitions between two-level systems in Fig. 3.4, which can be symbolically shown as

$$(\mathrm{g}) + h\nu \rightarrow (\mathrm{ex}) \rightarrow (\mathrm{ex})' + n\hbar\omega \rightarrow (\mathrm{g})' + h\nu' \rightarrow (\mathrm{g}) + n'\hbar\omega. \quad (3.50)$$

Here $h\nu$ denotes a photon and $\hbar\omega$ denotes a phonon. Actually, the process towards right is well observed in a photoabsorption process, which is followed by the relaxation and a photoemission of the material. It may seem to be evident that the system moves from left to right naturally in this way, similar to water flows from high to low level. However, according to the principle of the detailed balance in quantum mechanics, each transition probability from right to left should occur with an equal probability.

$$(\mathrm{g}) + h\nu \leftarrow (\mathrm{ex}) \leftarrow (\mathrm{ex})' + n\hbar\omega \leftarrow (\mathrm{g})' + h\nu' \leftarrow (\mathrm{g}) + n'\hbar\omega \quad (3.51)$$

Why not observe the leftward evolution? The reason is, it is easy to prepare one of $(\mathrm{g}) + h\nu$ states in (3.50) as an initial state, but it is practically impossible to prepare the right hand state $(\mathrm{g}) + n'\hbar\omega$ as an initial state which involves a large number of phonon states which must have exact phase correlations. In other words, we can easily create an outward spherical wave just by throwing a stone into a pond, but it is impossible to create an inward spherical wave by preparing the corresponding initial condition for the reverse process (Fig. 3.9).

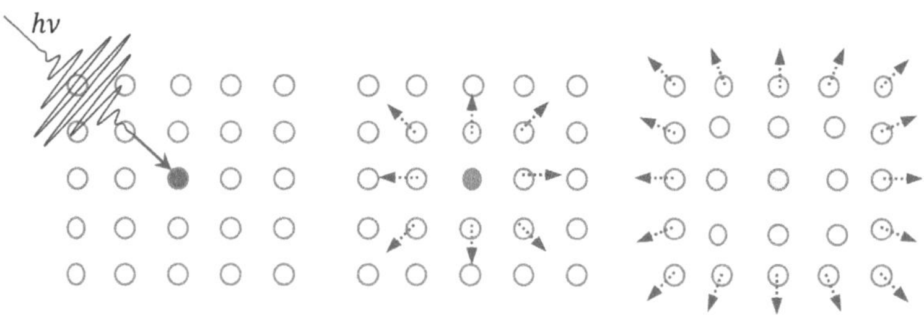

Figure 3.9 Lattice relaxation process (from left to right) after a photo excitation of an electron at a localized center. Impossible to create the reverse process (from right to left) by preparing the corresponding initial condition.

References

1. Shinozuka, Y., in *Materials and Reliability Handbook for Semiconductor Optical and Electron Devices*, ed. by Ueda, O., and Pearton, S., (Springer-Verlag, Berlin, 2012) Chapter 10.
2. Toyozawa, Y., and Inoue, M., *J. Phys. Soc. Jpn.*, **21** (1966) 1663.
3. Toyozawa, Y, *Optical Processes in Solids* (Cambridge Univ. Press, Cambridge, 2003).
4. Shinozuka, Y., in *Defect in Optoelectronic Materials*, ed. by Wada, K., and Pang, S. (Gordon and Bleach Publisher, Amsterdam, 1999) Chapter 8, p. 345.
5. Shinozuka, Y., Proc. MRS 1996 Fall Meeting **442**, *Defects in Electronic Materials II*, ed. Michel, J., Kennedy, T., Wada, K., and Thonke, K., p. 225.
6. Huang, K., and Rhys, A., *Proc. R. Soc. London A*, **204** (1950) 406.
7. Ueta, M., Kanzaki, H., Kobayashi, K., Toyozawa, Y., and Hanamura, E., *Excitonic Processes in Condensed Matter* (Springer, 1986).
8. Shinozuka, Y., *J. Phys. Soc. Jpn.*, **51** (1982) 2852.

Chapter 4

Localization: Self-Trapping

In Chapter 2, we have seen that the interaction between a carrier and the crystal phonons makes them a compound quasiparticle, polaron. The character of a polaron is classified into two types: large polaron and small polaron. A large polaron is realized when the electron–lattice (e–L) interaction is weak, where the wave function of an electron extends over large area and accompanied lattice distortion is weak. When the electron–lattice interaction is strong, a small polaron is realized where the extension of the electron wave function is small and accompanied lattice distortion is strong. In this chapter, we further discuss self-trapping of a carrier in semiconductors in two different approaches: One is the effective mass approximation with a continuum model and the other is a tight binding approach with a discrete model. Finally, we discuss the photoabsorption and photoemission processes in connection with self-trapping.

4.1 Study of Polaron (Self-Trapping) with Continuum Model

In this section, we will study how the character of a polaron depends on the strength of the electron–lattice interaction under

Electron–Lattice Interactions in Semiconductors
Yuzo Shinozuka

ISBN 978-981-4800-96-9 (Hardcover), 978-1-003-02346-3 (eBook)
www.jennystanford.com

the adiabatic approximation. Here, the electron motion is treated by the effective mass approximation and the lattice is regarded as a superposition of the elastic medium and the dielectric medium, respectively representing the acoustic phonon field and the optical phonon field. In contrast to the original analysis by Toyozawa [1], we present here a somewhat different variational approach to investigate the electron–lattice ground state.

4.1.1 Perfect Crystal

Consider an electron (hole) in the conduction (valence) band, whose effective mass is m^*, interacts with the bulk crystal lattice vibrations (acoustic and optical). The total Hamiltonian is given by

$$h_{\mathrm{e}}^{\mathrm{eff}} + V_{\mathrm{eL}} + U_{\mathrm{L}} = -\frac{\hbar^2}{2m^*}\nabla^2 + \sum_{\sigma,\boldsymbol{k}} A_{\sigma,\boldsymbol{k}}^{\mathrm{c}}(\boldsymbol{r})\, q_{\sigma,\boldsymbol{k}} + \frac{1}{2}\sum_{\sigma,\boldsymbol{k}} \omega_{\sigma,\boldsymbol{k}}^2 q_{\sigma,\boldsymbol{k}}^2. \tag{4.1}$$

Here, the kinetic energy ($\propto \dot{q}_{\sigma,\boldsymbol{k}}{}^2$) of the lattice is omitted since we will take the adiabatic approximation. The index σ refers the phonon mode σ = ac, op, and we will use the standing wave type modes for the wave vector $\boldsymbol{k}$, which is suitable for a localized electron: the present phonon modes correspond to $(q_{\sigma,\boldsymbol{k}} \pm q_{\sigma,-\boldsymbol{k}})/\sqrt{2}$ in Chapter 1.

Acoustic phonon

$$\omega_{\mathrm{ac},\boldsymbol{k}} = uk,$$

$$A_{\mathrm{ac},\boldsymbol{k}}^{\mathrm{c}}(\boldsymbol{r}) = -\left(\frac{2}{NM}\right)^{1/2} E_{\mathrm{d}}\, k \times \begin{pmatrix} \cos(\boldsymbol{kr}) & \text{for } k_x > 0 \\ \sin(\boldsymbol{kr}) & \text{for } k_x < 0 \end{pmatrix}, \tag{4.2}$$

where u is the sound velocity, E_{d} the deformation potential, M the mass of a unit cell, and N the number of unit cells. The above sine and cosine k-dependence comes from the standing-wave type representation: cf. Eq. (2.30).

Optical phonon

$$\omega_{\mathrm{op},\boldsymbol{k}} = \omega_{\mathrm{op}},$$

$$A_{\mathrm{op},\boldsymbol{k}}^{\mathrm{c}}(r) = -\left(\frac{2\omega_{\mathrm{op}}^2 e^2}{V\epsilon\varepsilon_0}\right)^{1/2} \frac{1}{k} \times \begin{pmatrix} \cos(\boldsymbol{kr}) & \text{for } k_x > 0 \\ \sin(\boldsymbol{kr}) & \text{for } k_x < 0 \end{pmatrix}, \tag{4.3}$$

where ω_{op} is the angular frequency of the optical phonon, V the volume of the crystal, e the electron charge, and $\epsilon = (\epsilon_\infty{}^{-1} - \epsilon_s{}^{-1})^{-1}$. Here ϵ_0 and ϵ_∞ are the static and high frequency dielectric constants, respectively: cf. Eq. (2.37).

In the variational method, the total energy of the system is a functional of the electron wave function $\psi(\boldsymbol{r})$ and the lattice displacements $\{q_{\sigma,\boldsymbol{k}}\}$:

$$E(\psi,\{q_{\sigma,\boldsymbol{k}}\}) = \langle\psi(\boldsymbol{r})|h_{\text{e}}^{\text{eff}} + V_{\text{eL}}+U_{\text{L}}|\psi(\boldsymbol{r})\rangle. \tag{4.4}$$

A standard procedure is to minimize first with respect to $\psi(r)$ for a fixed lattice configuration $\{q_{\sigma,\boldsymbol{k}}\}$: correspond to solve the Schrödinger equation. Next, minimize the so-called adiabatic potential with respect to the lattice displacements $\{q_{\sigma,\boldsymbol{k}}\}$. As far as the minimum energy of the system is concerned, which is expected to be obtained after both minimizations, we can change the order. We minimize $E(\psi\{q_{\sigma,\boldsymbol{k}}\})$ first with respect to $\{q_{\sigma,\boldsymbol{k}}\}$ for a fixed $\psi(\boldsymbol{r})$, and then minimize the result with respect to $\psi(\boldsymbol{r})$.

Suppose an electron is localized around a certain lattice point $\bar{\boldsymbol{R}}$ in a semiconductor with a normalized spherical wave function,

$$\psi_a(\boldsymbol{r}-\bar{\boldsymbol{R}}) = \left(\frac{\sqrt{2}}{a}\right)^{\frac{3}{2}} \exp\left[-\pi\frac{(\boldsymbol{r}-\bar{\boldsymbol{R}})^2}{a^2}\right], \tag{4.5}$$

where a is the extension of the wave function and is assumed to be a variational parameter. It is noted that this choice of a trial wave function breaks the translational symmetry in the original Hamiltonian. Since one can choose the center of localization at any $\bar{\boldsymbol{R}}$'s, then the translational symmetry would recover if we take a linear combination of the resulting $\psi_a(\boldsymbol{r}-\bar{\boldsymbol{R}})$'s, which has little effect compared to that without linear combination: see the discussion in Holstein model in Section 4.5.

The expectation value of the Hamiltonian with respect to the trial function (4.5) at a certain point, for example $\bar{\boldsymbol{R}} = 0$ is given by

$$E(\psi,\{q_{\sigma,\boldsymbol{k}}\}) = \left\langle\psi_a(\boldsymbol{r})\left|-\frac{\hbar^2}{2m^*}\nabla^2\right|\psi_a(\boldsymbol{r})\right\rangle - \sum_{\sigma,\boldsymbol{k}} q_{\sigma,\boldsymbol{k}}\langle\psi_a(\boldsymbol{r})|\,A_{\sigma,\boldsymbol{k}}^{\text{c}}(\boldsymbol{r})\,|\psi_a(\boldsymbol{r})\rangle + \frac{1}{2}\sum_{\sigma,\boldsymbol{k}} \omega_{\sigma,\boldsymbol{k}}^2 q_{\sigma,\boldsymbol{k}}^2. \tag{4.6}$$

Let each lattice mode $q_{\sigma,\boldsymbol{k}}$ be displaced from zero to respective equilibrium position, $\bar{q}_{\sigma,\boldsymbol{k}} = \langle\psi_a(\boldsymbol{r})|\,A^{\mathrm{c}}_{\sigma,\boldsymbol{k}}(\boldsymbol{r})\,|\psi_a(\boldsymbol{r})\rangle/A\omega^2_{\sigma,\boldsymbol{k}}$, which is a functional of the trial electron wave function. After integration with the trial wave function (4.5), the resulting total lattice relaxation energy $E^{\sigma}_{\mathrm{LR}}(a) = \frac{1}{2}\sum_{\boldsymbol{k}} \omega^2_{\sigma,\boldsymbol{k}}\bar{q}^2_{\sigma,\boldsymbol{k}}$ for the σ mode is found to be

$$E^{\mathrm{ac}}_{\mathrm{LR}}(a) = \frac{1}{2}\sum_{k_x>0}\sum_{k_y}\sum_{k_z} \frac{\left(\frac{2}{NM}\right) {E_d}^2 k^2}{(uk)^2} \exp\left(-\frac{a^2k^2}{4\pi}\right), \tag{4.7}$$

$$E^{\mathrm{op}}_{\mathrm{LR}}(a) = \frac{1}{2}\sum_{k_x>0}\sum_{k_y}\sum_{k_z} \frac{\frac{2\omega^2_{\mathrm{op}}e^2}{V\epsilon\varepsilon_0}}{\omega^2_{\mathrm{op}}k^2} \exp\left(-\frac{a^2k^2}{4\pi}\right). \tag{4.8}$$

In principle, the wave vector $\boldsymbol{k}$ should be restricted within the first Brillouin zone, then we have to introduce a cut-off vector in the above summation, such as the Debye cut off wave vector (Eq. (1.59)). But here, we will apply an alternative cut-off procedure where the variational parameter a is supposed to be limited in the range $0 \le d/a \le 1$ where d is the lattice constant, because the wave function is an envelope function, and the extension of the wave function should not shrink to less than d. Under this assumption, the summation over (k_x, k_y, k_z) can be replaced by a spherical integration from 0 to infinity. Adding the increase of an electron kinetic energy $\langle\psi_a(\boldsymbol{r})|(-\hbar^2/2m^*)\nabla^2|\psi_a(\boldsymbol{r})\rangle$, the total energy of the system is finally shown to be

$$\begin{aligned} E(a) &= \frac{3\pi\hbar^2}{2m^*a^2} - \frac{E^2_{\mathrm{d}}}{2Ca^3} - \frac{e^2}{4\pi\epsilon\varepsilon_0 a} \\ &= B \times \left(\frac{d}{a}\right)^2 - E^{\mathrm{ac}}_{\mathrm{LR}} \times \left(\frac{d}{a}\right)^3 - E^{\mathrm{op}}_{\mathrm{LR}} \times \left(\frac{d}{a}\right), \end{aligned} \tag{4.9}$$

where $B \equiv \dfrac{3\pi\hbar^2}{2m^*d^2}$, $\quad E^{\mathrm{ac}}_{\mathrm{LR}} \equiv \dfrac{E^2_{\mathrm{d}}}{2Cd^3}$, $\quad E^{\mathrm{op}}_{\mathrm{LR}} \equiv \dfrac{\mathrm{e}^2}{4\pi\epsilon\varepsilon_0 d}$.

In the last form (4.9), the variational parameter is switched to the inverse of the normalized extension d/a. The parameters characterize the material are: the strength of electron kinetic energy $B = 3\pi\hbar^2/2m^*d^2$, the absolute strength of the electron–

lattice interaction $E_{\mathrm{LR}}^{\mathrm{ac}} = E_{\mathrm{d}}^{2}/2Cd^{3}$ and $E_{\mathrm{LR}}^{\mathrm{op}} = e^{2}/4\pi\epsilon\varepsilon_{0}d$. The same result is also derived by use of the continuum model by Toyozawa [1].

Figure 4.1 shows the d/a dependence of the total energy $E(a)$ of the system. It is readily seen that the behavior of $E(a)$ around $d/a \sim 0$ reflects the difference in the force range of the e–L interactions. We will further search the true minimum in the range $0 \le d/a \le 1$. The realized value of $\overline{d/a}$ that gives the minimum of $E(a)$ depends on the relative strength of the electron–lattice interaction: $\vec{g} \equiv \vec{E}_{\mathrm{LR}}/B \equiv (E_{\mathrm{LR}}^{\mathrm{ac}}/B,\ E_{\mathrm{LR}}^{\mathrm{op}}/B) \equiv (g_{\mathrm{ac}}, g_{\mathrm{op}})$.

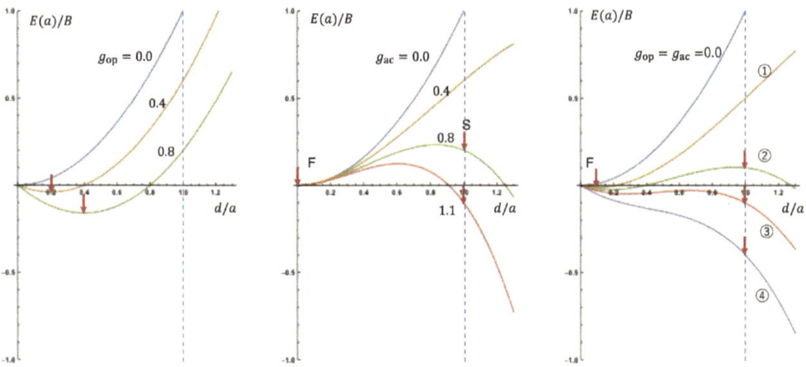

(a) optical mode (b) acoustic mode (c) optical mode + acoustic mode

Figure 4.1 The expectation value of the total energy of the electron-lattice system as a function of the inverse of the normalized extension $0 \le d/a \le 1$. Numbers 1–4 in (c) correspond to the parameter in the phase diagram in Fig. 4.2.

(a) Optical mode alone

There is always one minimum. As the strength g_{op} increases, $\overline{d/a}$ increases continuously from zero to 1. Here, $\bar{a}$ corresponds to the Pekar-type polaron radius [2] and gradually shrinks from ∞ to d.

(b) Acoustic mode alone

When the strength g_{ac} is weak, there is only one minimum at $\overline{d/a} = 0$, which corresponds to a free state without lattice distortion ($\bar{q}_{\sigma,\boldsymbol{k}} = 0$). As g_{ac} increases, there appears a strongly localized state $\overline{d/a} = 1$ with large lattice distortion ($\bar{q}_{\sigma,\boldsymbol{k}} \neq 0$) at first as a metastable state ($2/3 < g_{\mathrm{ac}} < 1$) and then as a stable state ($g_{\mathrm{ac}} > 1$).

There is always a potential barrier which separates from the free state ($\bar{a} = \infty$) and the strongly localized (self-trapped) state ($\bar{a} = d$).

(c) Optical mode + Acoustic mode

The extension of the stable state is either $\bar{a} \sim \infty$ or $\bar{a} = d$ depending on the parameter.

Figure 4.2 shows the phase diagram of the ground state of an electron in the phonon fields. In the region **F**, the extension of the electron wave function is large ($\bar{a} \gg d$) with weak lattice distortion; thus a large polaron is realized, where an electron almost freely moves through the crystal. In the region **S**, an electron is strongly localized ($\bar{a} = d$) with large lattice distortion; thus a small polaron is stable, where an electron is trapped by self-induced lattice distortion at some point in the crystal: we call self-trapped electron. If there is a metastable state, it is indicated in the parenthesis. It should be noted that there is a discontinuous transition line (thick solid) in the phase diagram. If the material parameter crosses the solid line, the character of a polaron discontinuously changes between **F** and **S**. The character of relaxed electronic excitations (conduction electron, valence hole, exciton, ...) in non-metallic materials falls in one of the regions in this phase diagram. Details will be discussed in the next section.

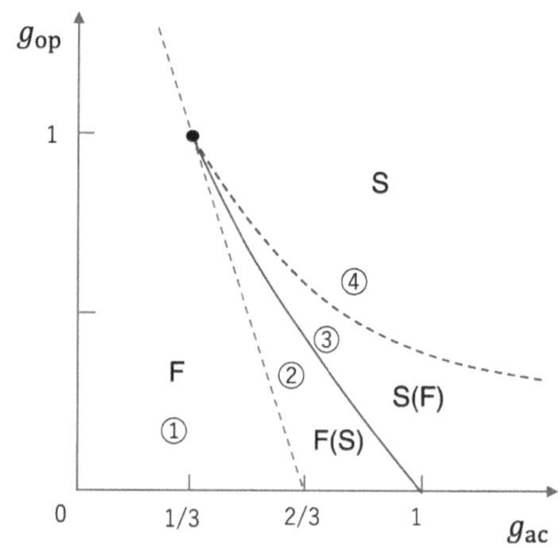

Figure 4.2 The phase diagram of an electron interacting with the lattice both by a short-range type (g_{ac}) and a long-range type (g_{op}). Number 1~4 correspond to the parameter in Fig. 4.1(c).

4.1.2 Characterization of Carriers in Non-Metallic Material

Let us discuss the character of various carriers in crystal semiconductors with the aid of the phase diagram given in Fig. 4.2. Based on experimental results, electrons, holes, and excitons in typical insulators and semiconductors are allocated in an appropriate point in the phase diagram in Fig. 4.3, where the character of an (electron, hole, exciton) is displayed with respect to the relative strength of the (short range, long range) part of the electron–lattice coupling $(E_{\mathrm{LR}}^{\mathrm{ac}}/B,\ E_{\mathrm{LR}}^{\mathrm{op}}/B) = (g_{\mathrm{ac}}, g_{\mathrm{op}})$ [1, 3]. Since an exciton is electronically neutral, their parameters are located on the $g_{\mathrm{op}} = 0$ axis. In a real system, excitons are classified into the Frenkel-type and the Wannier excitons. Even for the Wannier type, if the hole–lattice (h–L) interaction is dominant, the character of the relaxed exciton is governed by that of the relaxed hole, and then the present results can be applied.

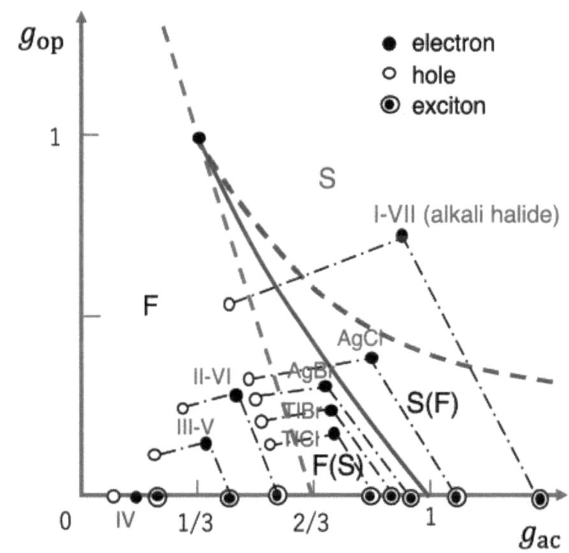

Figure 4.3 Electrons, holes and excitons in typical insulators and semiconductors allocated in appropriate region in the phase diagram of Fig. 4.2. After [3] with a minor revision.

In group IV semiconductors (C, Si, Ge), which have no iconicity in covalent bonding, the electron–lattice interactions are short-range type for both the acoustic and optical phonons. (Note that silicon crystal has two Si atoms in each unit cell; hence it has optical phonon modes, but the interaction is short range). Thus, the electron, hole, and exciton in Si are all located on the horizontal axis. Their effective masses are light; hence, B is large. Then the parameters are located on **F** region in the phase diagram. This result guarantees that carriers in IV materials can be used for electronic transport devices and serve for the semiconductor technology.

In III–V and II–VI semiconductors, the iconicity in bonding increases and also increases the long-range interaction by the optical phonon modes. But the strength is not enough to reach **S** region in the phase diagram. Then most carriers in III–V and II–VI semiconductors move almost freely in crystals. For a typical III–V semiconductor, parameters are $m^* = 0.5m_e$ $E_d = 6$ eV, $C = 4 \times 10^{10}$ N/m^2, $\epsilon = 4$ and $d = 3$Å, $B = 8.0$ eV, $E_{LR}^{op} = 1.2$ eV and $E_{LR}^{ac} = 2.7$ eV, for example. The resulting lattice distortion accompanied with carriers can be negligibly small (guarantee their application to electronic transport devices); however, these facts do not lead to a conclusion that strong electron–lattice phenomena are absent in these semiconductors. Several experimental phenomena are explained in terms of extrinsic self-trapping in the next section.

In I–VII materials, i.e., alkali halides, the long-range Fröhlich interaction becomes so strong for a hole to be located in the **S** region in the diagram. This hole self-trapping causes an exciton self-trapping in spite of electrons still in the **F** region because of its light effective mass. Self-trapped excitons in several alkali halides show symmetry breaking structural change, which will be discussed in Chapter 8.

4.1.3 Isoelectronic Impurity: Extrinsic Self-Trapping

Imperfections in crystals, such as an impurity atom, lattice vacancy, lattice dislocation, etc., affect electronic states. The most influential effect would be a carrier localization. For example, in silicon crystal, when a silicon atom is replaced by a phosphor atom, a localized state appears just below the conduction band edge, called a donor level, which is usually occupied by an extra conduction electron. Next let us consider the effect of e–L interaction to a conduction electron in the presence of a point defect.

Adding a spherical static defect potential $V_{\mathrm{der}}(r)$ to the Hamiltonian (4.1), the total energy, a functional of the wave function $\varphi(r)$ expressed as

$$E(\varphi,\{q_{\sigma,k}\}) = \langle \varphi(r)|h_{\mathrm{e}}^{\mathrm{eff}} + V_{\mathrm{eL}} + U_{\mathrm{L}} + V_{\mathrm{def}}(r)|\varphi(r)\rangle. \quad (4.10)$$

Here we assume that the phonon modes are not affected by the presence of a defect: possible effects will be discussed later (Chapter 6). The defect potential can be further decomposed into a long-range part and a short-range part as $V_{\mathrm{def}}(r) = V_l(r) + V_s(r)$ (Fig. 4.4), which are assumed to have forms

$$\text{the long-range part } V_l(r) = -V_l\, d/r, \quad (4.11\text{a})$$

$$\text{the short-range part } V_s(r) = \begin{cases} -V_s = \text{const.} & 0 \le r \le d \\ 0 & d \le r \end{cases}, \quad (4.11\text{b})$$

In case of a donor impurity, a Coulomb type potential can be employed for the long-range part with $V_l = e^2/4\pi\epsilon_\infty\varepsilon_0 d$: It should be careful that the high frequency ϵ_∞ is used here in order not to double count the dielectricity originated from the interaction with phonons. For an isoelectronic impurity, the constant V_s in the short-range part can be estimated by the difference in the electron affinity between host and impurity.

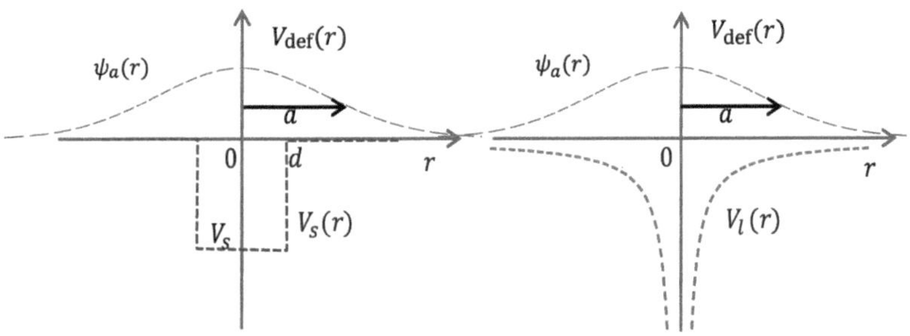

Figure 4.4 Typical profile of defect potential $V_{\mathrm{def}}(r)$ for the long range part and the short range part, where d is the lattice constant.

First, we extremalize with respect to the lattice $\{q_{\sigma,k}\}$. The resulting expectation value of the total energy for the trial function (4.5) is approximately given by

$$E(d/a) = B(d/a)^2 - (V_s + E_{\mathrm{LR}}^{\mathrm{ac}})\,(d/a)^3 - (V_l + E_{\mathrm{LR}}^{\mathrm{op}})\,(d/a). \quad (4.12)$$

We have neglected higher order terms of $(d/a)^n$: $n \geq 4$. The stable electron radius $\bar{a}$ now depends on two vector strengths for electron localization: the intrinsic parameter $\vec{E}_{\rm LR}/B \equiv (E^{\rm ac}_{\rm LR}/B,\ E^{\rm op}_{\rm LR}/B)$ and the extrinsic parameter $\vec{V}_{\rm def}/B \equiv (V_s/B, V_l/B)$.

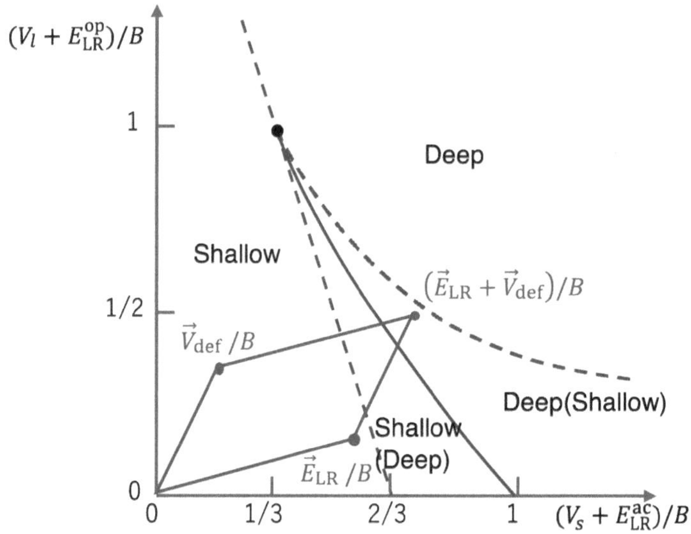

Figure 4.5 The phase diagram of an electron interacting with a point defect and the lattice with the short range type $(V_s + E^{\rm ac}_{\rm LR})$ and the log range type $V_l + E^{\rm op}_{\rm LR})$.

After a similar calculation in Section 4.1.1, we obtain the phase diagram of an electron (carrier) interacting with the lattice and a point defect (Fig. 4.5). In the region **Shallow**, an electron is shallowly localized around a point defect ($\bar{a} \gg d$) with small lattice distortion. In the region **Deep**, it is deeply localized ($\bar{a} = d$) with large lattice distortion. If there is a metastable state, it is indicated in the parenthesis. For a typical donor in nonpolar semiconductors such as P in Si, where $E^{\rm op}_{\rm LR} \sim 0$, $V_s \sim 0$, $\epsilon_\infty >> 1$ and m is light, then $\bar{a} \gg d$ and $E^{\rm ac}_{\rm LR}(\bar{a}) \sim 0$ in spite of finite $E^{\rm ac}_{\rm LR}$. As a result, the effect of the electron–lattice interaction can be negligible for shallow donor (acceptor) impurities in Si. On the other hand, it is possible that even if both $\vec{E}_{\rm LR}/B$ and $\vec{V}_{\rm def}/B$ locate in the region **Shallow** but the vector sum $(\vec{E}_{\rm LR} + \vec{V}_{\rm def})$ locates in the region **Deep** or **Deep (Shallow)** as is shown in Fig. 4.5. Then the ground state is a deep state with large lattice distortion.

Here, a strong electron-localization is a trigger and also a result of a large lattice distortion. We have shown here that strong electron-localization, triggered by the short-range electron–lattice interaction (see a dependence in Eq. (4.3)), induces a shallow-deep instability at point defects in semiconductors. This mechanism is called the extrinsic self-trapping [4]. A nitrogen substitutional impurity in Si [5] might be an example of this extrinsic self-trapping (**Deep** (**Shallow**)), which accompanies a metastable shallow state with small lattice distortion. Other examples in experiments will be discussed later in Section 4.4.

4.1.4 Exciton (an Electron–Hole Pair) in the Phonon-Fields

An elementary electronic excitation in semiconductors is an electron–hole pair, exciton. There are two types of exciton depending on materials:

(a) Frenkel exciton: The distance of an electron and a hole of an exciton is small. The long-range Fröhlich-type interactions for an electron and a hole are canceled with each other. The center of mass motion is the only freedom. Then a Frenkel exciton can be treated as a single particle which interacts with acoustic phonons with deformation potential type.

(b) Wannier exciton: The average distance between an electron and a hole is large compared to the lattice constant. Then we must consider various possibilities in the relaxed states. The adiabatic Hamiltonian of the system is (see Eq. (2.52))

$$h_{\mathrm{e}}^{\mathrm{eff}} + h_{\mathrm{h}}^{\mathrm{eff}} + U_{\mathrm{L}} + V_{\mathrm{eL}} + V_{\mathrm{hL}} + V_{\mathrm{eh}}, \tag{4.13}$$

$$h_{\mathrm{e}}^{\mathrm{eff}} = \frac{\boldsymbol{p}_{\mathrm{e}}^2}{2m_{\mathrm{e}}^*}, \quad h_{\mathrm{h}}^{\mathrm{eff}} = \frac{\boldsymbol{p}_{\mathrm{h}}^2}{2m_{\mathrm{h}}^*}, \tag{4.14a}$$

$$U_{\mathrm{L}} = \frac{1}{2}\sum_{\sigma,\boldsymbol{k}} \omega_{\sigma,\boldsymbol{k}}^2 q_{\sigma,\boldsymbol{k}}^2, \tag{4.14b}$$

$$V_{\mathrm{eL}} = -\sum_{\sigma,\boldsymbol{k}} A_{\sigma,\boldsymbol{k}}^{\mathrm{c}}\, q_{\sigma,\boldsymbol{k}} \exp(i\,\boldsymbol{k}\,\boldsymbol{r}_{\mathrm{e}}), \quad \sigma = \mathrm{ac, op} \tag{4.14c}$$

$$V_{\mathrm{hL}} = -\sum_{\sigma,\boldsymbol{k}} A^{\mathrm{v}}_{\sigma,\boldsymbol{k}}\, q_{\sigma,\boldsymbol{k}} \exp(i\, \boldsymbol{k}\boldsymbol{r}_{\mathrm{h}}), \qquad \sigma = \mathrm{ac, op} \tag{4.14d}$$

$$V_{\mathrm{eh}} = \frac{e^2}{4\pi\epsilon_\infty\varepsilon_0}\frac{1}{|\boldsymbol{r}_{\mathrm{e}} - \boldsymbol{r}_{\mathrm{h}}|}. \tag{4.14e}$$

Note that the dielectric constant in Eq. (4.14e) is the high frequency ϵ_∞.

Sumi studied the ground state of an electron–hole pair in the acoustic phonon field with deformation potential type interaction using a variational method [6]. Figure 4.6 shows the phase diagram on the stable state of an electron–hole pair with the Coulomb attraction in the acoustic phonon field with deformation potential interaction. Here $\mathrm{G_e} \equiv 2g_{\mathrm{e}}$ and $\mathrm{G_h} = 2g_{\mathrm{h}}$ are the coupling constant of deformation potential interations. In the region **Fex**, an exciton moves almost freely through the crystal. In the region **Se:h** (**Sh:e**), an electron (a hole) is self-trapped and a hole (an elecron) is loosly bound around it. In the region **Se:Sh**, an electron and a hole are self-trapped separately at different but near sites with opposite induced lattice distortions. This interesting configuration is realized in off-center type self-trapped exciton in alkali halides (see Section 4.4.1). In the region **Sex**, an electron and a hole are tigltly bound and self-trapped at the same site. The last case is caused by the cooperation of e–L and h–L interactions.

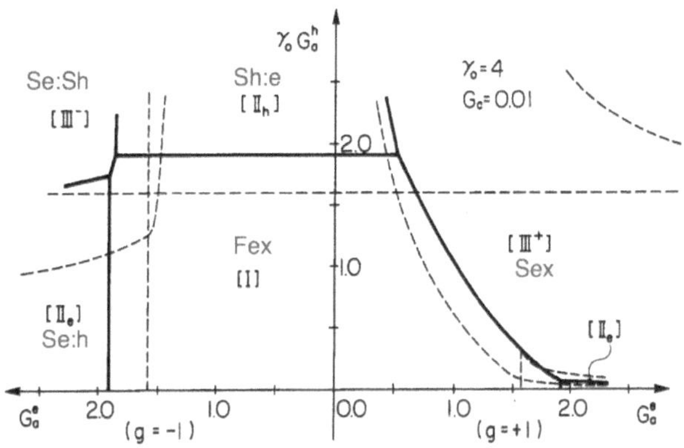

Figure 4.6 The phase diagram for the stable state of an electron–hole pair with the Coulomb attraction in the acoustic phonon field with the deformation potential interaction. After [6] with a minor revision.

Shimamura and Matsuura studied on the exciton self-trapping with a more detailed wave function [7]. Self-trapping of a Wannier exciton will be discussed on symmetry breaking later in this Chapter and also in Chapter 8.

4.2 Study of Self-Trapping Using Tight Binding Model

So far, we have learned that the trigger of self-trapping is a short-range type electron–lattice interaction. It would be informative to study self-trapping in various situations with a unified scheme paying attention to the short-range interaction using tight binding models. Although here we will refer to an electron in the conduction band, the following arguments are applicable for a hole in the valence band and a Frenkel exciton.

The Hamiltonian we consider hereafter is

$$H \equiv h_{\rm e} + V_{\rm eL} + U_{\rm L} = \sum_{l} |\phi_l\rangle\, \varepsilon_l \langle\phi_l| + \sum_{l}\sum_{l'} |\phi_l\rangle\, t_{ll'} \langle\phi_{l'}|$$

$$+ \sum_{l} |\phi_l\rangle\, (-\beta Q_l) \langle\phi_l| + \frac{1}{2}\sum_{l} C\, Q_l^2. \tag{4.15}$$

The system consists of an array of sites defined by a lattice vector $\boldsymbol{T}_l$ (Fig. 4.7). Each site has an atomic (Wannier) orbital $\phi_l(\boldsymbol{r})$ and a local lattice distortion Q_l with an elastic constant C. We use bra and ket presentation. Here ε_l is the site diagonal energy (depends on the site l in general) of an electron which interacts with the lattice distortion Q_l by a short-range-type interaction with a coupling constant β, which corresponds to the deformation potential constant $E_{\rm d}$ in the model in Section 4.1. An electron moves in the system with the electron transfer energy $t_{ll'}$ between $|\phi_l\rangle$ and $|\phi_{l'}\rangle$. The kinetic energy of the lattice is omitted since we will take the adiabatic approximation. This Hamiltonian can be applicable for various situations including, pure crystal (Section 4.2.2), low-dimensional system (Section 4.2.3), random system such as alloy or amorphous, confined system as quantum well (Section 4.2.4), and impurity (defect) system (Section 6.1.2). The energy spectrum of $h_{\rm e}$ is given in Appendix V.

4.2.1 Sum Rule on the Lowest (Relaxed) States

Our task is to find out the lowest (relaxed) state of the Hamiltonian (4.15) allowing the lattice to deform and to check the possibility of self-trapping. This can be done by calculating the adiabatic potential $U(\{Q_l\})$ as a function of $\{Q_l\}$ and find its minimum. As for the extremum (minimum or maximum) $\{\bar{Q}_l\}$, there is a following useful sum rule [4]:

$$\frac{\partial U(\{Q_l\})}{\partial Q_l} = \frac{\partial \langle\psi|H|\psi\rangle}{\partial Q_l} = \left\langle\psi\left|\frac{\partial H}{\partial Q_l}\right|\psi\right\rangle$$
$$= \langle\psi|[|\phi_l\rangle(-\beta)\langle\phi_l| + C\, Q_l]|\psi\rangle$$
$$= -\beta|c_l|^2 + C\, Q_l. \tag{4.16}$$

At the extremum $\partial U(\{Q_l\})/\partial Q_l = 0$, we then have

$$\bar{Q}_l = \left(\frac{\beta}{C}\right)|c_l|^2, \quad \left(\frac{C}{\beta}\right)\sum_l \bar{Q}_l = 1, \tag{4.17}$$

where $|\psi\rangle = \Sigma_l c_l|\phi_l\rangle$. This rule holds for any form of the electron part in the Hamiltonian, i.e., either for impurity or alloy, and helps us to figure out a gross picture of a free state and self-trapped state (Fig. 4.7). For example,

(a) free state in pure systems: ε_l = const.

The wave function is extended as much as possible over the crystal and the lattice distortion is negligibly small.

$$|\psi_{k=0}\rangle = (\sqrt{N})^{-1}\sum_l |\phi_l\rangle, \qquad \bar{Q}_l = \beta/NC \sim 0 \text{ for all } l. \tag{4.18}$$

$$E_{\text{free}} \sim \varepsilon - B,\ t_k = \frac{1}{N}\sum_{l,l'} t_{ll'} e^{ik(T_l - T_{l'})}, 2B = \text{Max}[t_k] - \text{Min}[t_k]$$

Here N is the number of total sites, ε is constant site energy.

(b) self-trapped state:

The wave function is localized at a particular site l, accompanied by a strong lattice distortion at l.

$$|\psi\rangle \sim |\phi_l\rangle, \quad \bar{Q}_l \sim \beta/C, \quad \bar{Q}_{l'(\neq l)} \sim 0, \quad E_{st} \sim \varepsilon - E_{LR}, \tag{4.19}$$

where $E_{LR} \equiv \beta^2/2C$ is the maximum lattice relaxation energy for a completely localized electron.

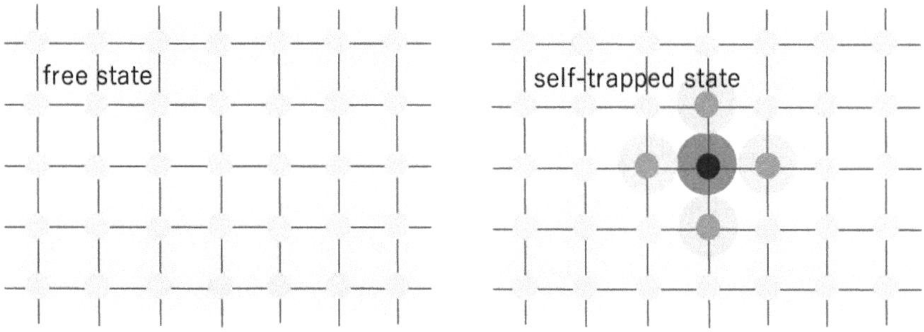

Figure 4.7 Illustration of a free state and self-trapped state. The color density of each circle indicates schematically the amplitude of the trial wave function: Eq. (4.22).

4.2.2 Homogeneous Pure System

First, we examine homogeneous pure systems consisting of orthorhombic $(2J_x + 1) \times (2J_y + 1) \times (2J_z + 1)$ lattice sites, where J_i is an integer (Fig. 4.8) [8]. The site $\boldsymbol{T}_i$'s are arrayed along $\boldsymbol{l} = (l_x, l_y, l_z)$,

$$\begin{aligned} l_x &= -J_x, -J_x + 1, \ldots, -1, 0, 1, \ldots, J_x - 1, J_x \\ l_y &= -J_y, -J_y + 1, \ldots, -1, 0, 1, \ldots, J_y - 1, J_y \\ l_z &= -J_z, -J_z + 1, \ldots, -1, 0, 1, \ldots, J_z - 1, J_z \, . \end{aligned} \tag{4.20}$$

The site energy ε in the Hamiltonian (4.15) is common for all sites, and hence can be set as $\varepsilon = 0$. The transfer energy is assumed to be restricted to nearest neighbors:

$$\begin{aligned} &t_{ll'} = -t_x(< 0) \text{ for } l_x = l'_x \pm 1, \ l_y = l'_y, \quad l_z = l'_z \\ &t_{ll'} = -t_y(< 0) \text{ for } l_x = l'_x, \quad l_y = l'_y \pm 1, \ l_z = l'_z \\ &t_{ll'} = -t_y(< 0) \text{ for } l_x = l'_x, \quad l_y = l'_y, \quad l_z = l'_z \pm 1. \\ &\text{Otherwise } t_{ll'} = 0 \end{aligned} \tag{4.21}$$

The model describes various situations, including

(1) infinite isotropic system

exact 3-dimension $J_x = l_y = l_z = \infty$
exact 2-dimension $J_x = l_y = \infty$, $l_z = 0$
exact 1-dimension $J_x = \infty$, $l_y = l_z = 0$

(2) infinite anisotropic system

quasi 2-dimension $t_x = t_y \gg t_z$
quasi 1-dimension $t_x = t_y \ll t_z$

(3) quantum well of infinite potential barrier

one-dimensional QW (slab) $J_x = l_y = \infty$, l_z is finite
two-dimensional QW (wire) $J_x = \infty$, l_y and l_z are finite
three-dimensional QW (dot) J_x, l_y and l_z are finite.

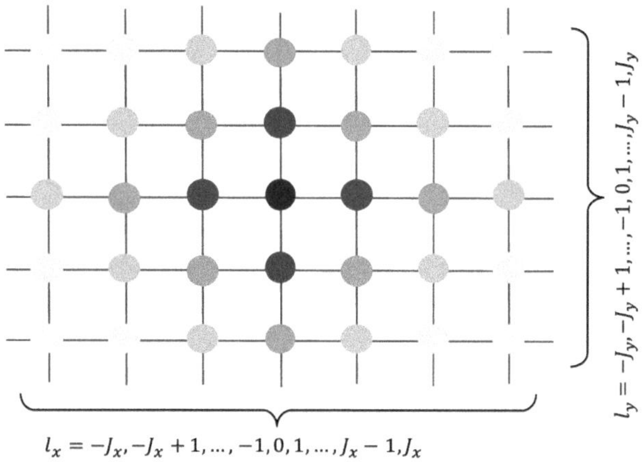

Figure 4.8 An example of $(2J_x + 1) \times (2J_y + 1) \times (2J_z + 1)$ array model with $J_x = 3$, $J_y = 2$, and $J_z = 0$. The color density of each circle indicates schematically the amplitude of the trial wave function (4.22).

Hereafter we search for the minimum of the adiabatic potential $U(\{Q_l\}) = \langle \psi | H | \psi \rangle$, where Q_l is a local lattice distortion at site $\boldsymbol{T}_l$, with a following trial function

$$|\psi\rangle = c_0 \sum_{l_x=-J_x}^{J_x} \sum_{l_y=-y}^{J_y} \sum_{l_z=-J_z}^{J_z} \exp\left[-\kappa_x |l_x| - \kappa_y |l_y| - \kappa_z |l_z|\right] |\phi_{T_l}\rangle. \tag{4.22}$$

Here c_0 is a normalization constant and κ_i's (($i = x, y, z$)) are variational parameters: $\kappa_i = 0$ corresponds to an extended state along the i-th direction, while a set of large κ_i's corresponds to a localized state at the origin. The expecting value of the Hamiltonian (4.15) for this trial function (4.22), which has been extremalized with respect to $\{Q_l\}$ is shown to be

$$U(\kappa_x, \kappa_y, \kappa_z) = \varepsilon - \sum_{i=x,y,z} 4t_i\, \xi_{J_i}(\kappa_i) - E_{\mathrm{LR}} \prod_{i=x,y,z} \eta_{J_i}(\kappa_i), \quad (4.23)$$

where

$$\xi_{J_i}(\kappa_i) = \frac{\exp(-\kappa_i)\,\{1 - \exp(-2J_i\kappa_i)\}}{1 + \exp(-2\kappa_i) - 2\exp(-2(J_i + 1)\kappa_i)}, \quad (4.24a)$$

$$\eta_{J_i}(\kappa_i) = \frac{\{1 - \exp(-2\kappa_i)\}\{1 + \exp(-4\kappa_i) - 2\exp(-4(J_i + 1)\kappa_i)\}}{\{1 + \exp(-2\kappa_i)\}\{1 + \exp(-2\kappa_i) - 2\exp(-2(J_i + 1)\kappa_i)\}^2}. \quad (4.24b)$$

Here $E_{\mathrm{LR}} = \beta^2/2C$. As have done in the previous section, we will minimize $U(\kappa_x, \kappa_y, \kappa_z)$ with respect to κ_i's. The results are as follows [8].

(a) infinite isotropic system

In the rigid lattice $\{Q_l = 0\}$, the electron eigenstates and energies are

$$|\psi_k\rangle = \frac{1}{\sqrt{N}} \sum_l \exp(i\boldsymbol{k} \cdot \boldsymbol{T}_l)\, |\phi_l\rangle, \quad (4.25)$$

$$\varepsilon_k = \varepsilon + 2t\,\{\cos(k_x a) + \cos(k_y a) + \cos(k_z a)\} \quad \text{for 3 - dim.} \quad (4.26a)$$

$$\varepsilon_k = \varepsilon + 2t\,\{\cos(k_x a) + \cos(k_y a)\} \quad \text{for 2 - dim.} \quad (4.26b)$$

$$\varepsilon_k = \varepsilon + 2t\cos(k_x a) \quad \text{for 1 - dim.} \quad (4.26c)$$

Here a is the lattice constant and the full width of the electron energy band is $2B = 4t \times \delta$ (δ-dimension: $\delta = 1, 2, 3$) centered at ε. Hence the lowest energy is $\varepsilon - B$ at $\boldsymbol{k} = 0$.

Figure 4.9 shows the total energy U as a function of the localization index $\alpha_i \equiv 1 - \exp(-\kappa_i)$, which varies $0 \le \alpha_i \le 1$. The index $\alpha_i = 0$ corresponds to an extended state along the i-th direction, and $\alpha_x = \alpha_y = \alpha_z = 1$ corresponds to a completely localized state at the center. Here the coupling constant g is defined as

$$g \equiv E_{LR}/B, \quad \text{with} \quad B = 2t \times \delta \quad (\delta\text{-dimension}). \tag{4.27}$$

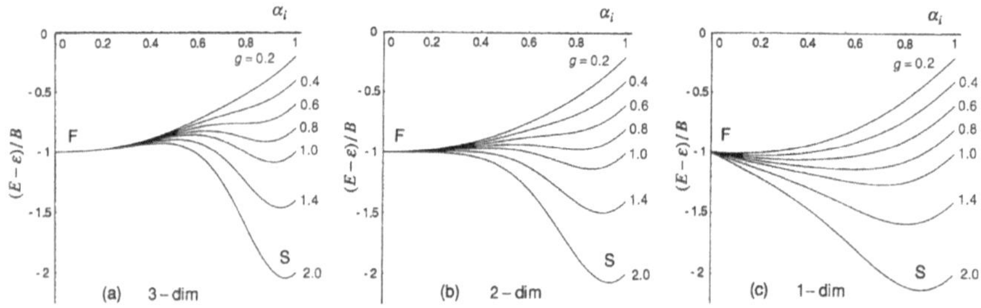

Figure 4.9 The total energy $U(=E)$ as a function of the localization index $\alpha_i\,(i = x, y, z)$ for exact (a) three-, (b) two- and (c) one-dimension. Reproduced with permission from Ref. [8].

It is found self-trapping occurs approximately when $g \geq 1$. To find the exact critical coupling constant, one needs to include a finite energy gain in the self-trapped state by an electron transfer to neighboring sites.

(a) 3-dim

Extended **F** state ($\alpha_x = \alpha_y = \alpha_z = 0$) is always stable or at least metastable with respect to distortions. As the coupling constant $g \equiv E_{LR}/B$ increases, **S**-state ($\alpha_x = \alpha_y = \alpha_z \sim 1$) first appears as metastable for $g \geq 0.605$ and then becomes stable for $g \geq 0.906$; the **F-S** transition is the first order. A potential barrier always exists which separates **F** and **S**.

(b) 2-dim

Extended **F** state ($\alpha_x = \alpha_y = 0$) is at first stable for $g \leq 0.706$. Self-trapped state **S** appears first as metastable for $0.706 \leq g \leq 0.840$ and then becomes stable for $0.840 \leq g$. Finally, **F** becomes unstable for $g > 2.00$.

(c) 1-dim

Extended state ($\alpha_x = 0$) is unstable for nonzero coupling constant g. The ground state changes its character from **F** to **S** continuously as g increases.

4.2.3 Infinite Anisotropic System

In discussing infinite anisotropic systems, it would be sufficient to study $t_x = t_y \neq t_z$ case and introduce the degree of anisotropy,

$$\lambda = t_x/t_z, \tag{4.28}$$

which varies between 0 and ∞. If we compare the present tight binding model with the effective mass approximation, it can be related with $\lambda \equiv m_z/m_x$, where m_i is the electron effective mass along the i-th direction.

$\lambda \ll 1$ corresponds to quasi one-dimensional (chain-like) systems.

$\lambda \gg 1$corresponds to quasi two-dimensional (layered) systems.

$\lambda = 1$ to the isotropic three-dimensional systems.

Figure 4.10 shows (a) the phase diagram of the ground state of an electron with respect to the anisotropy λ and the coupling constant g and (b) the anisotropy dependence of the energies of the **F** state, the **S** state and the potential barrier height for $g = 1$ case. It is found that the character of **F-S** transition is qualitatively the same as in isotropic three-dimension, and there always exists a potential barrier between **F** and **S**. But The barrier height decreases as the anisotropy increases ($\lambda \to 0, \infty$).

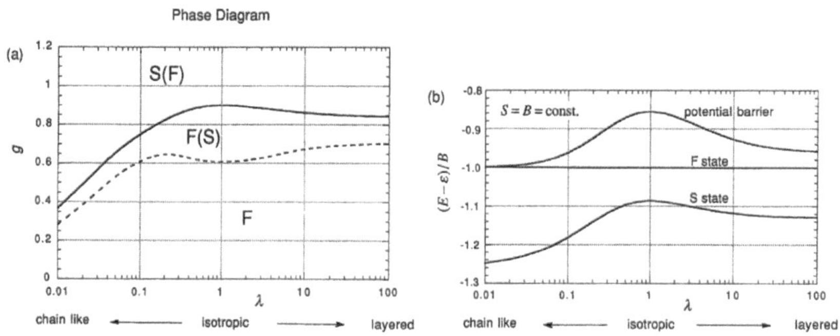

Figure 4.10 (a) The phase diagram of a relaxed electron with respect to the anisotropy λ and the coupling constant g. (b) the anisotropy dependence of the energies of the **F** state, the **S** state and the potential barrier height for $S(=E_{LR}) = B$ constant ($g = 1$) case. Reproduced with permission from Ref. [8].

4.2.4 Quantum Well

Here, we discuss the effect of quantum confinement to self-trapping, only in a case of isotropic transfer ($t_x = t_y = t_z = t$), for simplicity. In order to avoid any confusion, we will name here as follows: d-dimensional quantum well means that an electron is confined with respect to d-dimension, where it moves free in the rest 3-d dimension.

(a) one-dimensional QW (slab):

Figure 4.11a shows the phase diagram of an electron in one-dimensional QW (slab) with respect to the well size J_z and the e–L coupling $g \equiv E_{LR}/B$ ($B = 6t$) is the half band width in perfect 3-dim.). The character of F-S transition is qualitatively the same as in isotropic 3-dim., but as the well width becomes thin, i.e., when $J_z = 0$ it behaves as in the exact 2-dim. described above.

(b) two-dimensional QW (wire):

Figure 4.11b shows the phase diagram of an electron in two-dimensional QW (wire) with respect to the well size J_z and the e–L coupling $g \equiv E_{LR}/B$ ($B = 6t$) is the half band width in perfect 3-dim.). The character of F-S transition is qualitatively the same as in isotropic 3-dimension, but as the well width becomes thin, i.e., when $J_z = 0$, it behaves as in the exact 1-dim., where the transition occurs continuously.

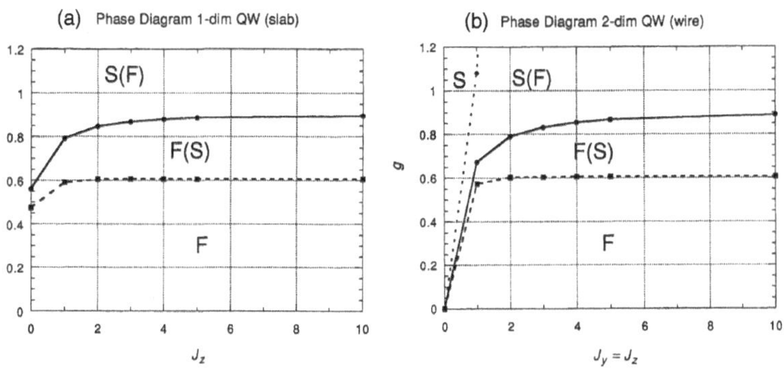

Figure 4.11 (a) The phase diagram of a ground state of an electron in deformable lattice of (a) 1-dim. QW (slab) and (b) 2-dim. QW (wire), with respect to the well size and the coupling constant g. A metastable state, if it exists, is shown in the parentheses. Reproduced with permission from Ref. [8].

4.3 Adiabatic Potential Using Tight Binding Model

So far, we have studied the lowest energy state of an electron (hole, exciton) in the deformable lattice. In order to discuss optical absorption and emission processes related with exciton self-trapping, it would be helpful to figure out the overall adiabatic potential of the system, where Hamiltonian (4.15) is applied to an exciton, i.e., Frenkel exciton and also to a Wannier exciton if one of the interaction with an electron or a hole is dominant: usually a hole is the main trigger for exciton self-trapping because of heavy effective mass.

Let us recall a simple tight binding model, where the Hamiltonian is given by Eq. (4.15)

$$H \equiv h_e + V_{eL} + U_L$$
$$= \sum_l |\phi_l\rangle \varepsilon_l \langle\phi_l| + \sum_l \sum_{l'} |\phi_l\rangle\, t_{ll'} \langle\phi_{l'}| - \sum_l |\phi_l\rangle \beta Q_l \langle\phi_l| + \frac{1}{2}\sum_l CQ_l^2 . \tag{4.29}$$

The site diagonal energy ε_l's are common for all sites in pure crystals, then we can set $\varepsilon_l = \varepsilon = 0$. We have seen in Section 3.2 that the interaction mode $Q_{\rm int}$ plays a key role in the lattice relaxation processes. The interaction mode which connects with F state and S state can be approximated by a local distortion Q_l at arbitral site l because of the sum rule (4.17) and the fact $(C/\beta)\bar{Q}_l \sim 1$. Thus, our task is to calculate the electron energies as a function of one of Q_l. Then without loss of generality, we focus on $l = 0$, then $Q_{\rm int} \sim Q_0$. Neglecting other modes, the Hamiltonian is simplified as

$$H = |\phi_0\rangle(-\beta Q_0)\langle\phi_0| + \sum_l \sum_{l'} |\phi_l\rangle\, t_{ll'} \langle\phi_{l'}| + \frac{1}{2} CQ_0^2. \tag{4.29a}$$

Now the problem is reduced to a "point defect" problem where only the potential at $l = 0$ is lowered by βQ_0.

4.3.1 Adiabatic Potential in 3-dim. Crystals

The adiabatic potential $U(Q_0)$ can be calculated using the Green's function method given in Appendix VI: see Eqs. (A6.14–A6.16). Typical examples for $U(Q_0)$ are illustrated in Fig. 4.12. As the lattice distortion Q_0 increases from zero, no localized state appears at the beginning, because the short-range potential $-\beta Q_0$ is weak (see Appendix VI), then $U(Q_0)$ increases along the lattice elastic energy. When the potential $-\beta Q_0$ exceeds a critical value, a localized state appears below the main exciton band, and the corresponding adiabatic potential splits off. Along this potential branch, a local minimum (corresponds to a self-trapped state) appears first as a metastable state then becomes stable for $g \geq g_c$, where g is the coupling constant defined by Eq. (4.27). $g \equiv E_{LR}/B$ and $E_{LR} \equiv \beta^2/2C$.

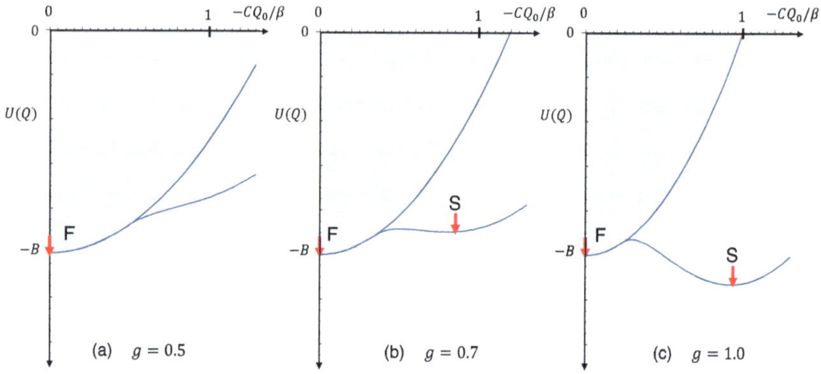

Figure 4.12 Typical examples of the adiabatic potentials as a function of the normalized distortion $-CQ_0/\beta$. Minimum at $Q_0 = 0$ corresponds to a free state F. Self-trapped state S at $-CQ_0/\beta \sim 1$ is (a) unstable, (b) metastable, and (c) stable.

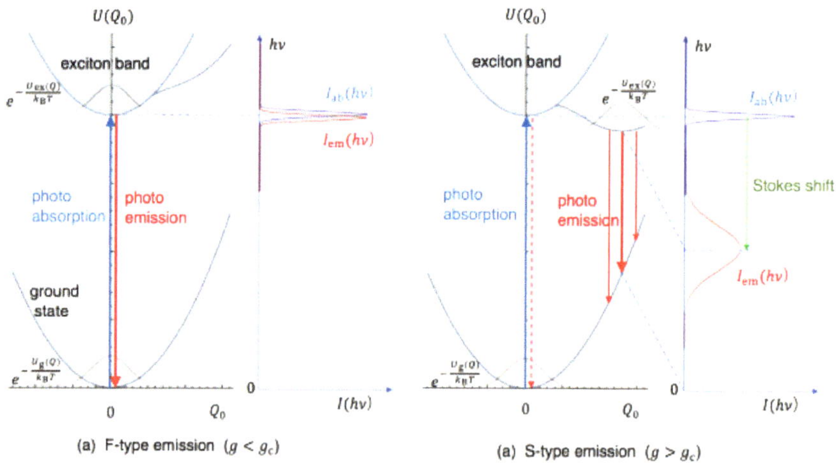

Figure 4.13 Adiabatic potentials and optical spectra: (a) sharp F-type emission in $g < g_c$. (b) broad S-type emission in $g < g_c$ with large Stokes shift.

After a calculation for particular transfer $t_{ll'}$ forms, it is found that

$g_c = 0.90$ for a nearest neighbor transfer model

$g_c = 0.84$ for a hemielliptic density of states model (Fig. 4.13).

The results are qualitatively the same as those obtained before $g_c \sim 1$. A small difference in the critical values of g_c comes from a difference in a finite energy gain due to the transfer energy gain in S state.

4.3.2 Optical Spectra

The optical spectra in exciton self-trapping can be discussed similarly to the two-level system given in Section 3.2.

(a) Absorption

Photoabsorption occurs from the ground state with no exciton where the lattice randomly vibrates around the equilibrium position $\{Q_l = 0\}$. At low temperature, the spectrum sharply rises at the exciton peak energy $\varepsilon - B$ for Frenkel exciton. For Wannier exciton, it locates just below the band gap edge. At high temperature, there arises a possibility of transient self-trapped exciton, which will be discussed in the next section.

(b) Emission

Photoemission occurs from the relaxed excited state: if F state is stable, the equilibrium positions of the lattice are $\{(C/\beta)\bar{Q}_l = 1/N \sim 0\}$, and the photoemission occurs around $Q_0 \sim 0$, the spectrum is sharp consisting of free exciton annihilation with a very small Stokes shift. If S state is stable, the equilibrium positions of the lattice are $\{(C/\beta)\bar{Q}_l \sim 1\}$ for a particular l and the others are $\{(C/\beta)\bar{Q}_{l' \neq l} \sim 1\}$. The photoemission occurs from the relaxed excited state around $Q_l \sim \bar{Q}_l$, then the spectrum is a broad Gaussian (width $\sim \sqrt{E_{LR} k_B T}$) whose peak is at $\varepsilon - 2E_{LR}$ ($E_{LR} \equiv \beta^2/2C$), resulting in a large Stokes shift $2E_{LR} - B$ (see Section 3.4.1) (Fig. 4.13).

4.3.3 Isoelectronic Impurity (Extrinsic Self-Trapping)

Here we consider a perfect 3-dimensional crystal where only at the origin ($\boldsymbol{T}_l = 0$) the host A atom is replaced by an Isoelectronic impurity of another B atom. Following a similar procedure done in Section 4.3.1, we will check the effect of a large distortion at the impurity site $\boldsymbol{T}_l = 0$. The chosen Hamiltonian is

$$H = |\phi_0\rangle(\Delta - \beta Q_0)\langle\phi_0| + \sum_l \sum_{l'} |\phi_l\rangle t_{ll'} \langle\phi_{l'}| + \frac{1}{2}\sum_l C\, Q_l^2. \quad (4.30)$$

Here $\Delta \equiv \varepsilon_B - \varepsilon_A$ is the site energy difference between the impurity B and the host A. We can set $\varepsilon_A = 0$ as the energy origin. Figure 4.14 shows typical adiabatic potentials $U(Q_0)$ and the obtained phase diagram [7]: A calculation procedure is given in Appendix VI.

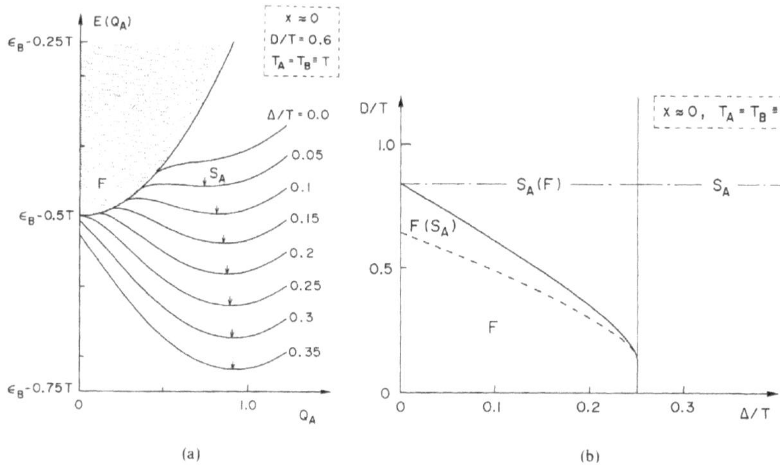

Figure 4.14 (a) Typical adiabatic potentials for an electron interacting with the lattice by short-range-type interactions at a point defect with potential depth $-\Delta$. (b) Phase diagram using $T \equiv 2B$, the full band width, and $D \equiv 2E_{LR}$, twice of the lattice relaxation energy. Reproduced with permission from Ref. [9].

When the effective impurity potential $(\Delta - \beta Q_0)$ is not so strong compared to the half band width B, it cannot bind an electron; hence no localized state appears. Then the lowest electronic energy is constant, $-B$, while the lattice elastic energy increases. When the potential $(\Delta - \beta Q_0)$ overcomes a critical value, a localized state appears below the main electron band, and the corresponding adiabatic potential splits off, the one-electron energy decreases roughly given by $(\Delta - \beta Q_0)$. Figure 4.14a shows several typical adiabatic potentials, and (b) is the phase diagram of an electron interacting with the lattice by a short-range type in the presence of a point defect for the nearest neighbor transfer model. In the triangular region of $\Delta/B < 0.68$ and $E_{\mathrm{LR}}/B < 0.9$, an electron is self-trapped even for neither of them alone is enough for localization. This phenomenon is called "extrinsic-self-trapping": the cooperation of e–L interactions and a static potential.

4.3.4 Alloy Semiconductor (Mixed Crystal)

Next, we will discuss self-trapping in a random system, i.e., substitutional alloy A_xB_{1-x}, which is composed of a regular array of sites $\{l\}$, each occupied either by an A atom or B atom with a probability x or $1 - x$, respectively. The Hamiltonian described these alloys can be presented as

$$H = \sum_l |\phi_l\rangle \varepsilon_l \langle\phi_l| + \sum_l \sum_{l'} |\phi_l\rangle t_{ll'} \langle\phi_{l'}| - \sum_l |\phi_l\rangle \beta_l Q_l \langle\phi_l| + \frac{1}{2}\sum_l C_l Q_l^2. \tag{4.31}$$

There are several possibilities in randomness

diagonal randomness $\varepsilon_{\mathrm{A}} \neq \varepsilon_{\mathrm{B}}, \beta_{\mathrm{A}} \neq \beta_{\mathrm{B}}, C_{\mathrm{A}} \neq C_{\mathrm{B}}$
site off-diagonal randomness $t_{\mathrm{AA}} \neq t_{\mathrm{BB}} \neq t_{\mathrm{AB}}$

The parameters are related to those previously introduced as

$T_\alpha = 2B_\alpha$ (=$12t_{\alpha\alpha}$ for simple cubic, nearest neighbor transfer)
$D_\alpha = 2E_{\mathrm{LR}}^\alpha = \beta_\alpha^2/C_\alpha$, for $\alpha = \mathrm{A, B}$.

We can evaluate the lowest state with following two candidates (Fig. 4.15):

(a) Free state:

The wave function is extended as much as possible over the alloy crystal and the lattice distortion is negligibly small.

$$|\psi_{k=0}\rangle = (\sqrt{N})^{-1}\sum_l |\phi_l\rangle, \qquad \bar{Q}_l = \frac{\beta_l}{NC_l} \sim 0 \text{ for all } l, \tag{4.32}$$

where N is the number of total sites. If the randomness is not so large, we use the virtual crystal approximation, where the energy of the free state is

$E_{\mathrm{free}} \sim x\varepsilon_{\mathrm{A}} + (1 - x)\varepsilon_B - B$ for site diagonal randomness

$E_{free} \sim \varepsilon - xB_A - (1 - x)B_B$ for site off-diagonal randomness

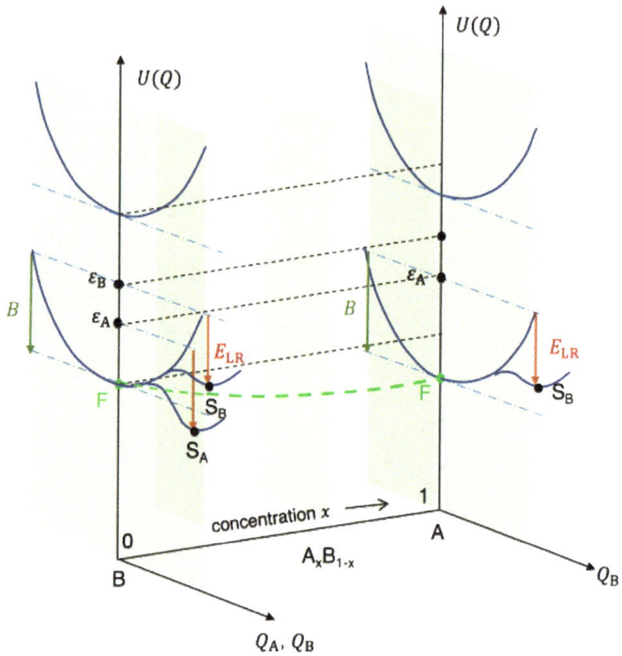

Figure 4.15 Illustration of CC diagrams of self-trapping in alloy semiconductors A_xB_{1-x}: $x = 0$ and 1.

(b) Self-trapped state:

The wave function is localized at a particular site, either at A-site or B-site, accompanied by a strong lattice distortion.

$$|\psi\rangle \sim |\phi_A\rangle,\ \bar{Q}_A \sim \beta_A/C_A,\ \bar{Q}_l \sim 0\ (l \neq A),\ \ E_{st,A} \sim \varepsilon_A - E_{LR}^{A} \quad (4.33a)$$

$$|\psi\rangle \sim |\phi_B\rangle,\ \bar{Q}_B \sim \beta_B/C_B,\ \bar{Q}_l \sim 0\ (l \neq B),\ \ E_{st,B} \sim \varepsilon_B - E_{LR}^{A} \quad (4.33b)$$

Since E_{free} depends on the alloy concentration x, then if $E_{st,A}$, or $E_{st,B}$ is lower than E_{free}, an electron is self-trapped, if not an electron extends through the alloy crystal. Figure 4.16 shows a typical example how the lowest state changes between S state and F-state as the concentration changes using the coherent potential approximation CPA [9]. In Section 4.4.3, we will discuss experimental results on several alloys with the present theory.

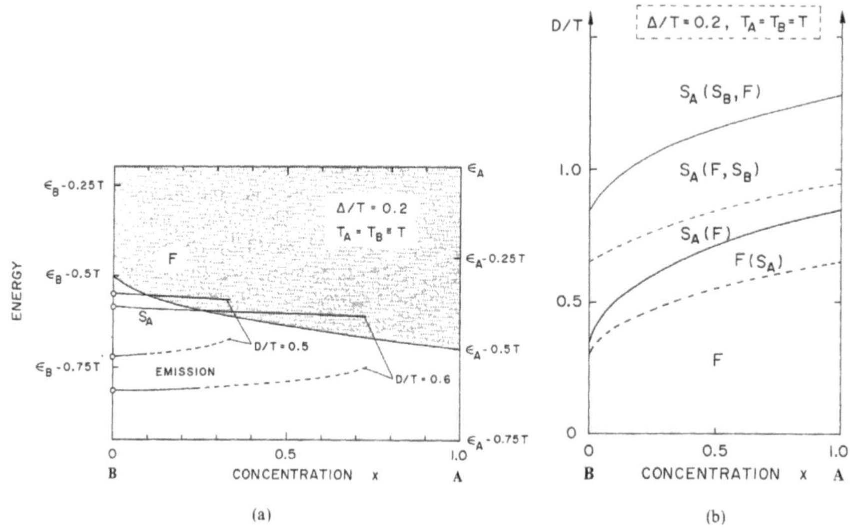

Figure 4.16 (a) Energies of F-state and S_A-state in an alloy semiconductor A_xB_{1-x}, with peak position of emission. The e–L coupling constant $g = D/T$ is 0.5 and 0.6, where S state is unstable in pure A crystal, but stable in B-rich alloys. (b) An example of the phase diagram for the potential difference Δ/T. Reproduced with permission from Ref. [9].

When the randomness is so large, there arise another possibility, i.e., cluster states where the wave function is extended within a cluster of A-sites and the lattice distortion is relatively small.

$$|\psi_{k=0}\rangle = (\sqrt{N_A})^{-1} \sum_l |\phi_A\rangle, \qquad \bar{Q}_A = \frac{\beta_A}{N_A C_A} \quad \text{for all } l, \tag{4.34}$$

where N_A is the number of total sites composed for the cluster. A detailed analysis is given in [4, 9].

4.4 Characterization of Carriers in Semiconductors

So far, we have learned that elementary electronic excitations (conduction electron, valence hole, exciton) in non-metallic materials (semiconductors, insulators) interact with phonons and can be qualitatively classified into a large polaron (F-type) or small polaron (S-type), depending on the strength of the interactions.

4.4.1 Alkali Halides and Ag Halides

Figure 4.17 shows photoabsorption and photoemission spectra of several alkali halide crystals. It is readily seen that all shows large Stokes shifts in emission spectra, indicating the relaxed exciton states are all S-type. There are two S-type emission bands showing middle and large Stokes shifts.

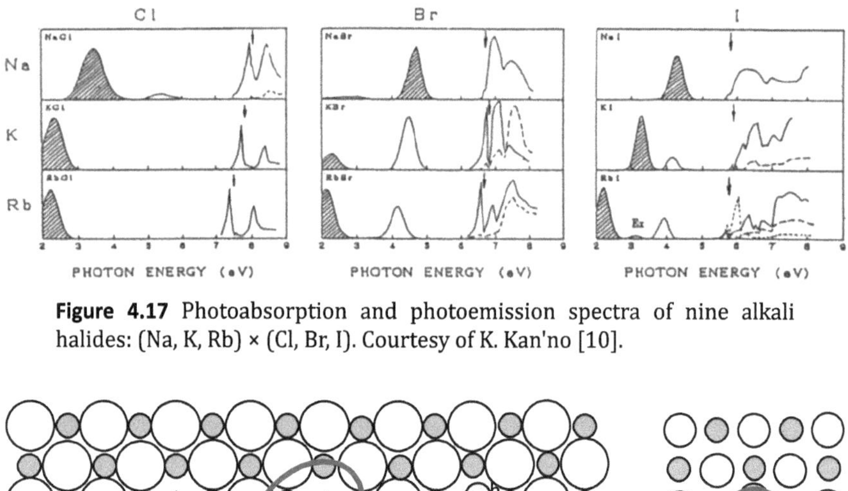

Figure 4.17 Photoabsorption and photoemission spectra of nine alkali halides: (Na, K, Rb) × (Cl, Br, I). Courtesy of K. Kan'no [10].

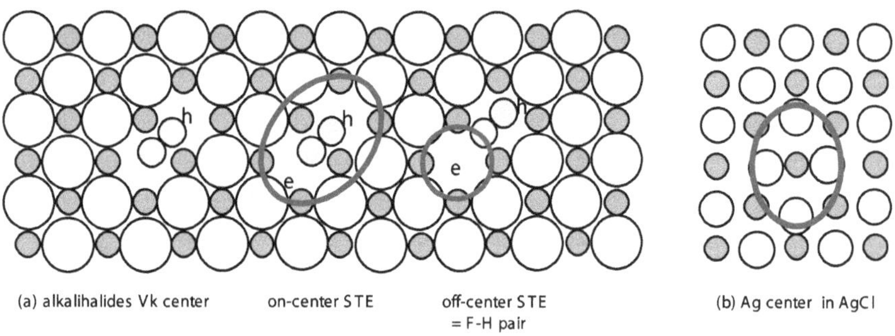

Figure 4.18 Illustration of the atomic configuration of STE in (a) alkali halides and (b) AgCl.

This experimental result is clearly explained by two possibilities in self-trapped exciton: on-center type and off-center type (Fig. 4.18). Figure 4.19 is an illustration of the CC diagram of two types. Q_1 denotes the attraction of two halogen ions relating the interaction mode between F-state and on center

STE (Vk-center+bound electron), Q_2 denotes their center of mass displacement indicating the interaction mode between on center and off-center STE. See details in Ref. [11].

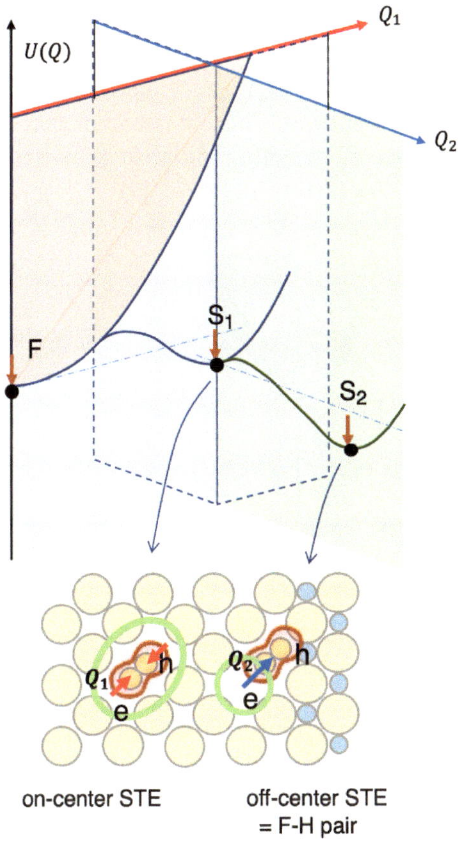

Figure 4.19 The configuration coordinate diagram in Q_1 and Q_2 modes for an exciton in alkali halides in 3-dimensional plot.

4.4.2 Impurity

We have learned in Section 4.1.3 a possibility of extrinsic self-trapping caused by the cooperation of e–L interaction and defect potential (Fig. 4.5). Then why does in Si the donor or the acceptor impurity not show any lattice relaxation effects, contrary to F-center in alkali halides, which show large lattice distortion. F-center in alkali halides is a cation vacancy, which effectively possesses +e charge and attracts a conduction electron. In Si, the long-range-type e–L interaction with optical phonons $E_{\text{LR}}^{\text{op}}$ is absent or weak, and the dielectric constant ϵ_∞ is large. Then even the cooperation with V_l and $E_{\text{LR}}^{\text{ac}}$ does not exceed large B due to small m^*, resulting in a large extension radius $\bar{a} \gg d$ and small lattice relaxation $\sim E_{\text{LR}}^{\text{ac}} \times (d/\bar{a})^3$. Unlike Si, in alkali halides, the long-range part of a cation vacancy potential V_l is strong due to small dielectric constant. Then the cooperation with large $E_{\text{LR}}^{\text{ac}}$ and V_l overcomes B, resulting in strong electron localization $\bar{a} \sim d$ and large lattice distortion (Fig. 4.20).

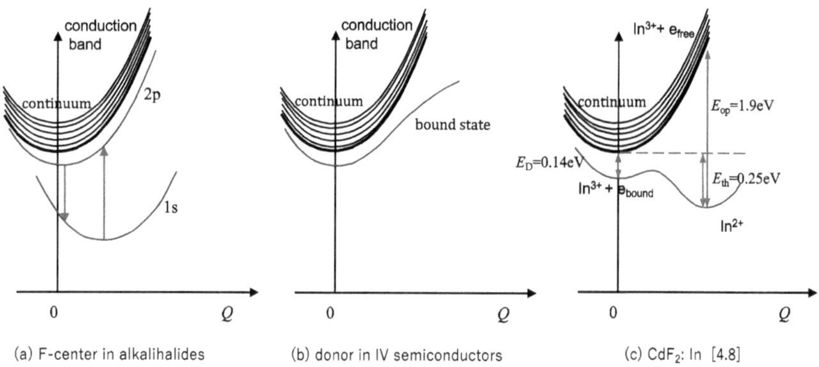

Figure 4.20 Typical adiabatic potentials for (a) F-center in alkali halides, (b) donor impurity in IV semiconductors, and (c) In impurity in CdF_2.

CdF_2 is an ionic semiconductor. Langer et al. [12] showed that a conduction electron trapped by +e In impurity shows bistable structures: one is shallowly localized state In^{3+}+e_{bound} with small lattice distortion and the other is deeply localized state In^{2+} with large lattice distortion. These phenomena can be explained if the parameter for CdF_2: In locates Deep(Shallow) region in the phase diagram in Fig. 4.5. Many defect systems in III-V semiconductors, such as DX-center and EL2 in GaAs also show large lattice distortion (see Chapter 8).

4.4.3 Alloy Semiconductor

Here we apply the above theoretical result on experimental results on substitutional alloy A_xB_{1-x}. In real alloys, all the physical parameters are affected to some degree by the disorder. From the comparison between the absorption edges of pure A- and B-crystals, we can only estimate the difference in $\varepsilon_A - T_A/2$ and $\varepsilon_B - T_B/2$ (see Eq. (4.31)). However, from the emission spectrum we can obtain much more information on the values of $\Delta \equiv \varepsilon_B - \varepsilon_A$, D_A and D_B: compare the results in Figs. 4.16(a) and (b), where $(\varepsilon_A - T_A/2) - (\varepsilon_B - T_B/2) = 0.2$ for all cases.

Relaxed exciton states in AgBr (=A) are F-states while those in AgCl (=B) are S-states localized at an Ag atom. In Fig. 4.19b we illustrate the distortion pattern, showing the Jahn–Teller effect (see Chapter 8). In terms of the model in Section 4.3.3, the parameters are estimated as $D_A/T_A < 0.84$, $D_B/T_B > 0.84$ and $\varepsilon_A < \varepsilon_B$. The appearance of Stokes shifted broad emission at $x \sim 0$ can be clearly explained. Figure 4.21 shows the photoabsorption spectrum and emission spectrum of $AgCl_{1-x}Br_x$ [13]. Although the absorption spectrum of $AgCl_{1-x}Br_x$ is the amalgamation type, the emission spectrum changes discontinuously from the S-type to the F-type at $x \sim 0.45$. This behavior is explained if we assume $T_A > T_B$, and agrees well with the results in Section 4.3.4. ESR experiments on the self-trapped states show that even in a Cl-rich concentration ($x < 0.45$), a hole is apt to be localized at an Ag-atom surrounded by more Br-atoms. If we regard a pair of an Ag atom and a Br (Cl) atom as an A(B) site, this is consistent with the result in Section 4.3.4 with $T_A > T_B$ where the energy of S_A-state is lower than that of S_B-states. This phenomenon can also be explained if we take account of a small perturbation to ε and/or D, that is, $\varepsilon_A < \varepsilon_B$ (see Fig. 4.16(b)) and/or $D_A > D_B$.

When $D_A/T_A > 0.84$ and $D_B/T_B > 0.84$, relaxed exciton states in mixed A_xB_{1-x} crystals are always S-states. This is the case in alkali halide mixed crystals, showing a two-center type self-trapping. The effect of alloying is not attractive in the context of F-S transition; however, the relative stability of S_A-states and S_B-states, which depends on the difference in each parameter (ε, β, D) gives us geometric information on self-trapped states in mixed and also pure crystals.

Excitons in $TlCl_{1-x}Br_x$, $SrS_{1-x}Se_x$, $ZnSe_{1-x}Te_x$ and $Zn_{1-x}Cd_xSe$ can be attributed to the case $\varepsilon_A \neq \varepsilon_B$. Although F-state is stable in both pure cases $x = 0$ and 1, the appearance of Stokes shifted broad emission at $x \sim 0$ is clearly explained in terms of extrinsic self-trapping shown in subsection 4.3.3. As x increases, gradual blue shift of the S-type emission and a sudden appearance of the F-type emission of $TlCl_{1-x}Br_x$ are also explained by the present results [14]. On the emission spectrum of $SrS_{1-x}Se_x$, however, as x increases the Stokes shifted S-type emission splits into two peaks and the high energy peak gradually changes into the F-type emission. This behavior cannot be explained by the present result and may be due to an extended state at a cluster of Se atoms: if D/T is rather small this state is stable in large clusters [4].

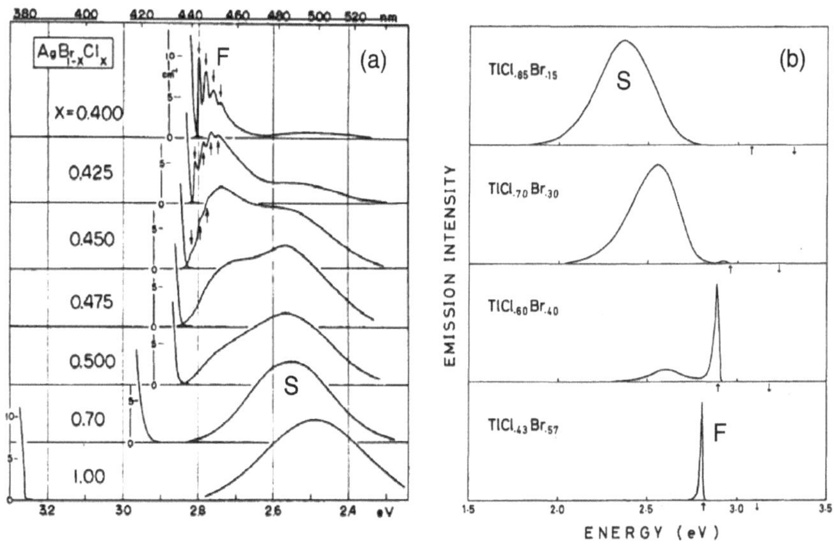

Figure 4.21 (a) photoabsorption and photoemission spectrum in $AgCl_{1-x}Br_x$. Reproduced with permission from Ref. [13]. (b) Photoemission in $TlCl_{1-x}Br_x$. As the concentration of Br increases, the spectrum discontinuously changes from **S**-type to **F**-type. Reproduced with permission from Ref. [14].

4.5 Small Polaron: Holstein Approach

So far, we have studied electron self-trapping phenomena in the adiabatic approximation, neglecting the effect of the kinetic energy of the lattice vibration. In order to discuss the effect, Holstein proposed a model [15], described by the following Hamiltonian:

$$H = \sum_{l}\sum_{l'} |\phi_l\rangle t_{ll'} \langle\phi_{l'}| - \sum_{l} |\phi_l\rangle \beta Q_l \langle\phi_l| + \frac{1}{2}\sum_{l} (P_l^2 + \omega^2 Q_l^2). \tag{4.35}$$

An electron moves through a one-dimensional lattice and couples with a localized phonon at each site (Fig. 4.22). Here the variables are used similar to Eq. (4.15), and Q_l and P_l are the lattice distortion and its conjugate momentum at the site l.

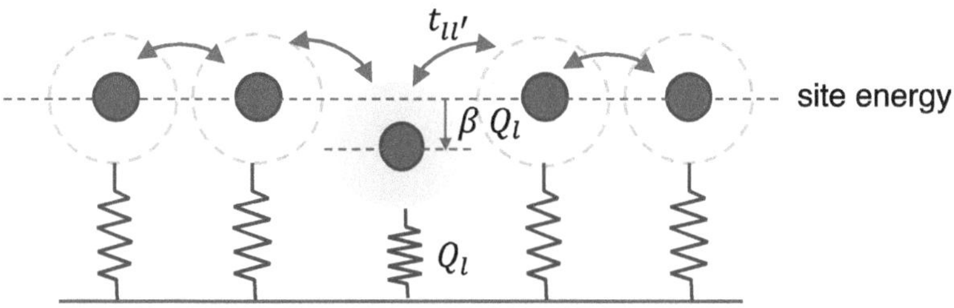

Figure 4.22 Holstein model [15] for small polaron.

If we neglect the electron transfer energy $t_{ll'}$, the system can be diagonalized, and the ground state is

$$|\phi_l\rangle \times |\Phi^{\mathrm{L}}_{l,n=0}\!: \bar{Q}_l = \beta/\omega^2\rangle, \tag{4.36}$$

where an electron is localized at site l, and the coupled phonon is in the ground state $n = 0$ with a shifted origin $\bar{Q}_l = \beta/\omega^2$. This is accompanied by the phonon ground state $n' = 0$ at all other sites l' with unshifted origin $\bar{Q}_{l'} = 0$.

When $t_{ll'} \neq 0$, let us try a linear combination of Eq. (4.36):

$$|\Psi_k\rangle = \frac{1}{\sqrt{N}}\sum_l \exp(i\boldsymbol{k}\cdot\boldsymbol{T}_l)\,|\phi_l\rangle \times |\Phi^{\mathrm{L}}_{l,n=0}\!: \bar{Q}_l = \beta/\omega^2\rangle. \tag{4.37}$$

The effective transfer energy between $\langle\phi_l| \times \langle\Phi^{\mathrm{L}}_{l,n=0}\!: \bar{Q}_l = \beta/\omega^2|$ and $|\phi_{l'}\rangle \times |\Phi^{\mathrm{L}}_{l',n'=0}\!: \bar{Q}_{l'} = \beta/\omega^2\rangle$ can be evaluated using Eq. (AII-7) in Appendix II. It contains,

$$\langle\Phi^{\mathrm{L}}_{l,n=0}| \times \langle\phi_l|H|\phi_{l'}\rangle \times |\Phi^{\mathrm{L}}_{l',n'=0}\rangle = t_{ll'}\,e^{-2S}, \tag{4.38}$$

where $S = E_{\mathrm{LR}}/\hbar\omega$ ($E_{\mathrm{LR}} = \beta^2/2\omega^2$) is the Huang–Rhys factor. Since the effective transfer energy of an electron is reduced by e^{-2S}, the effective mass of a polaron has the electron band mass multiplied by e^{2S}, which has an enormous value for a typical value $S > 5$, there this small polaron is so heavy and localized (self-trapped).

References

1. Toyozawa, Y., in *Proc. 4th Intern. Conf. Vacuum Ultraviolet Radiation Physics* (Vieweg-Pergamon, Braunschweig, 1974) p. 317.
2. Pekar, S., *Journ. Phys. U. S. S. R.*, **10** (1946) 341; **19** (1949) 796.
3. Toyozawa, Y, *Optical Processes in Solids* (Cambridge Univ. Press, Cambridge, 2003).
4. Shinozuka, Y., and Toyozawa, Y., *J. Phys. Soc. Jpn.*, **46** (1979) 505.
5. Brower, K. L., *Phys. Rev. B*, **26** (1982) 6040.
6. Sumi, A., *J. Phys. Soc. Jpn.*, **43** (1977) 1286.
7. Shimamura, S., and Matsuura, M., *Solid State Commun.*, **45** (1983) 547.
8. Shinozuka, Y., and Ishida, N., *J. Phys. Soc. Jpn.*, **64** (1995) 3007.
9. Shinozuka, Y., *J. Phys. Soc. Jpn.*, **59** (1990) 1322.
10. Kan'no, K., Tanaka, K., and Hayashi, T., *Rev. Solid State Sci.*, **4** (1990) 383.
11. Song, K. S., and Williams, R. T., *Self-Trapped Excitons* (Springer, Berlin, 1993).
12. Langer, J., Ogonowska, U., and Iller, A., *Inst. Phys. Conf. Ser.*, **43** (1979) 277.
13 Kanzaki, H., Sakuragi, S., and Sakamoto, K., *Solid State Commun.*, **9** (1971) 999.
14. Takahei, K., and Kobayashi, K., *J. Phys. Soc. Jpn.*, **44** (1978) 1850.
15. Holstein, T., *Ann. Phys.*, **8** (1959) 325 and 343.

Chapter 5

Optical Spectra in Perfect Crystals: Final State Interaction

Strong electron–lattice (e–L) interactions in semiconductors generate lattice distortion, carrier localization, structural change, etc., and also affect the optical transition processes. We have seen that the stable atomic configuration in the electronic ground state changes more or less when the electrons are excited. Then photoabsorption and photoemission processes reflect the character and strength of the electron–lattice interaction in the material [1]. The photoabsorption energy reflects the relative positions of the adiabatic potentials near the equilibrium in the electronic ground state. Thus, as well as photoemission, the photoabsorption spectra reflects the e–L interaction through the effects in the final state of the transition.

5.1 Adiabatic Instability of a Free Electron in a Deformable Lattice

In the previous chapter, we have learned that when the electron–lattice interaction is strong, a free electron (hole) in the conduction (valence) band becomes trapped by self-induced lattice distortion. This phenomenon, called self-trapping, is a

Electron–Lattice Interactions in Semiconductors
Yuzo Shinozuka

ISBN 978-981-4800-96-9 (Hardcover), 978-1-003-02346-3 (eBook)
www.jennystanford.com

consequence of the intrinsic instability due to the electron–lattice interaction. Whether self-trapping occurs or not depends on the interaction force range, its strength, and the dimensionality of the system. Before developing a general theory on the instability, it would be worth to study an elemental case with a simple tight binding model to investigate the adiabatic instability of the excited state.

Consider a tight binding model of a δ-dimensional (δ = 1, 2, 3) rectangular system whose volume is L^δ (L is the length) and the total site number is N^δ: ($d = L/N$ the lattice constant) [2]. An electron in the conduction band is assumed to interact with the lattice dilation through a short-range-type interaction. The adiabatic Hamiltonian we consider is

$$H = h_\mathrm{e} + V_\mathrm{eL} + U_\mathrm{L} = \sum_{\boldsymbol{n},\boldsymbol{m}} t_{\boldsymbol{nm}} |\boldsymbol{n}\rangle\langle\boldsymbol{m}| - \sum_{\boldsymbol{n}} \beta Q_{\boldsymbol{n}} |\boldsymbol{n}\rangle\langle\boldsymbol{n}| + \frac{1}{2}\sum_{\boldsymbol{n}} C\, Q_{\boldsymbol{n}}^2. \tag{5.1}$$

Here $|\boldsymbol{n}\rangle$ is an electron state localized at $\boldsymbol{n}$-th site, $Q_{\boldsymbol{n}}$ is the (normalized) lattice distortion with the elastic constant C, and β the coupling constant. In the Fourier forms, we have

$$|\boldsymbol{k}_\mathrm{e}\rangle = \frac{1}{\sqrt{N^\delta}}\sum_{\boldsymbol{n}} \exp\,[i\boldsymbol{k}_\mathrm{e}\boldsymbol{n}]\,|\boldsymbol{n}\rangle, \quad q_{\boldsymbol{k}} = \frac{1}{\sqrt{N^\delta}}\sum_{\boldsymbol{n}} \exp\,[i\boldsymbol{k}\boldsymbol{n}]\,Q_{\boldsymbol{n}}\,,$$

$$\varepsilon_{\boldsymbol{k}_\mathrm{e}} = \frac{1}{N^\delta}\sum_{\boldsymbol{n},\boldsymbol{m}} \exp\,[i\boldsymbol{k}_\mathrm{e}(\boldsymbol{n}-\boldsymbol{m})]\,t_{\boldsymbol{nm}} \approx \varepsilon_{\boldsymbol{k}_\mathrm{e}=0} + \frac{\hbar^2}{2m^*}{k_\mathrm{e}}^2 + \cdots\,,$$

$$H = \sum_{\boldsymbol{k}_\mathrm{e}} \varepsilon_{\boldsymbol{k}_\mathrm{e}} |\boldsymbol{k}_\mathrm{e}\rangle\langle\boldsymbol{k}_\mathrm{e}| - \frac{\beta}{\sqrt{N^\delta}}\sum_{\boldsymbol{k}_\mathrm{e}}\sum_{\boldsymbol{k}_\mathrm{e}'} q_{\boldsymbol{k}_\mathrm{e}-\boldsymbol{k}_\mathrm{e}'} |\boldsymbol{k}_\mathrm{e}\rangle\langle\boldsymbol{k}_\mathrm{e}'| + \frac{1}{2}\sum_{\boldsymbol{k}} C\, q_{\boldsymbol{k}} q_{-\boldsymbol{k}}. \tag{5.2}$$

Rigid lattice $\{Q_{\boldsymbol{n}} = 0\}$, $\{q_{\boldsymbol{k}} = 0\}$

The lowest electron state of h_e is $|\boldsymbol{k}_\mathrm{e} = 0\rangle$ which has an energy $\varepsilon_{\boldsymbol{k}_\mathrm{e}=0}$. Then the adiabatic potential starts from

$$U_{0\text{-th}}\,(q_{\boldsymbol{k}} = 0) = \varepsilon_{\boldsymbol{k}_\mathrm{e}=0}. \tag{5.3}$$

The deformable lattice $\{Q_n \neq 0\}, \{q_k \neq 0\}$

Once the lattice is allowed to deform, the electronic energy is lowered due to the e–L interaction V_{eL} while the elastic energy U_{eL} of the lattice increases on the other hand. Let us estimate the gain and loss taking V_{eL} as a perturbation.

First-order perturbation

The lowest state $|\boldsymbol{k} = 0\rangle$ directly couples with $q_{\boldsymbol{k}=0}$ mode:

$$U_{1\mathrm{st}}(q_{\boldsymbol{k}}) = \varepsilon_{\boldsymbol{k}_e=0} + \langle \boldsymbol{k}_e = 0|V_{\mathrm{eL}}|\boldsymbol{k}_e = 0\rangle + \frac{1}{2}\sum_{\boldsymbol{k}} C\, q_{\boldsymbol{k}} q_{-\boldsymbol{k}}$$
$$= \varepsilon_{\boldsymbol{k}_e=0} - \frac{\beta}{\sqrt{N^\delta}} q_{\boldsymbol{k}=\boldsymbol{0}} + \frac{1}{2}\sum_{\boldsymbol{k}} C\, q_{\boldsymbol{k}} q_{-\boldsymbol{k}}, \tag{5.4}$$

resulting the equilibrium distortion: $\bar{q}_{\boldsymbol{k}=\boldsymbol{0}} = \beta/C\sqrt{N^\delta}$ or $\bar{Q}_n = \beta/CN^\delta \sim 0$ for all sites. Thus, the effect is infinitesimal, and then can be neglected for macroscopic materials.

Second-order perturbation

Up to the second-order perturbation of V_{eL}, the result is

$$U_{2\mathrm{nd}}(q_{\boldsymbol{k}}) = \varepsilon_{\boldsymbol{k}_e=0} - \frac{\beta}{\sqrt{N^\delta}} q_{\boldsymbol{0}} + \frac{1}{2} C q_{\boldsymbol{0}} q_{\boldsymbol{0}}$$
$$+ \sum_{\boldsymbol{k}\neq\boldsymbol{0}} \left[\frac{\beta^2}{N^\delta} \frac{1}{\varepsilon_{k=0} - \varepsilon_k} + \frac{1}{2} C \right] q_{\boldsymbol{k}} q_{-\boldsymbol{k}}. \tag{5.5}$$

Among various distortion modes, the most effective and critical distortion would be $\boldsymbol{k}_1 = (2\pi/L, 0, 0)$ mode which gives the smallest denominator (Fig. 5.1). Putting $L = Nd$, the $q_{\boldsymbol{k}_1} q_{-\boldsymbol{k}_1}$ term has turned out to be

$$\left[-\frac{\beta^2 N^{2-\delta}}{4\pi^2 \left(\frac{\hbar^2}{2m^* d^2} \right)} + \frac{1}{2} C \right] q_{\boldsymbol{k}_1} q_{-\boldsymbol{k}_1}$$
$$= \frac{1}{2} C\left[1 - (E_{\mathrm{LR}}/2B) N^{2-\delta}\right] q_{\boldsymbol{k}_1} q_{-\boldsymbol{k}_1}. \tag{5.6}$$

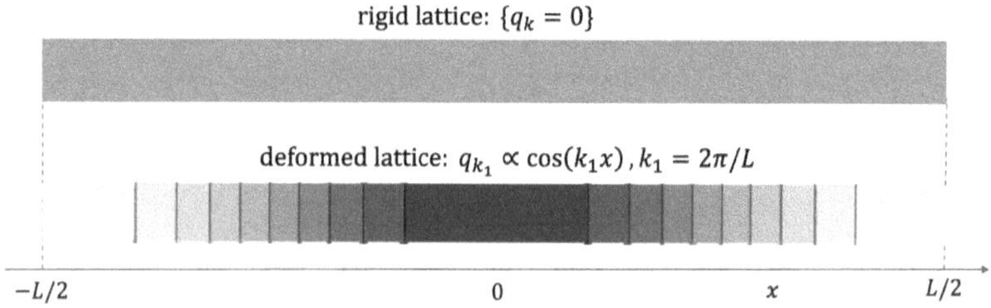

Figure 5.1 Contour plot image of effective potential induced by lattice distortion. A long wavelength lattice mode $q_{k'}$ affects the lowest energy electron state towards localization.

Here, we use parameters $B = \pi^2(\hbar^2/2m^*d^2)$ and $E_{\mathrm{LR}} = \beta^2/2C$.

(1) In one dimension (δ = 1), as $N^{2-\delta} \to \infty$ the coefficient always tends to $-\infty$, which implies that the free electron state is always unstable with respect to lattice distortions of long wavelength.

(2) In two dimension (δ = 2), the marginal case: the coefficient is positive or negative, depending on the sign of the bracket in (5.6). If $g \equiv E_{\mathrm{LR}}/B < 2$, the free electron state is stable. On the other hand, if $g \equiv E_{\mathrm{LR}}/B > 2$, the free electron state becomes unstable. This is consistent with the result obtained in Section 4.2.2 with a minor revision $\pi \cong 3$, originated from the difference in the distortion pattern.

(3) In three dimension (δ = 3), as $N^{2-\delta} \to 0$, the coefficient is always positive, this implies that the free electron state is always stable (at least metastable) with respect to any pattern (wavelength) of small lattice distortions. As the interaction becomes strong, on the other hand, there arises a self-trapped state with strong lattice distortions that is separated from the free state by a potential barrier.

The above results on the instability of a free electron state are consistent with the results obtained in Chapter 4.

Further analysis is given by Toyozawa and Shinozuka [2] for the general interaction Hamiltonian in δ-dimension. Let extend Eq. (5.2) as

$$H = \sum_{\boldsymbol{k}} \varepsilon_k |\boldsymbol{k}_{\mathrm{e}}\rangle\langle \boldsymbol{k}_{\mathrm{e}}| - \frac{1}{\sqrt{N^\delta}} \sum_{\sigma,\boldsymbol{k}_{\mathrm{e}}} \sum_{\boldsymbol{k}_{\mathrm{e}'}} A_{\sigma,\boldsymbol{k}_{\mathrm{e}}-\boldsymbol{k}_{\mathrm{e}'}} q_{\boldsymbol{k}_{\mathrm{e}}-\boldsymbol{k}_{\mathrm{e}'}} |\boldsymbol{k}_{\mathrm{e}}\rangle\langle \boldsymbol{k}_{\mathrm{e}}{}'| + \frac{1}{2}\sum_{\sigma,\boldsymbol{k}} \omega_{\sigma,\boldsymbol{k}}^2 \, q_{\boldsymbol{k}} q_{-\boldsymbol{k}}, \tag{5.7}$$

where the wave vector k-dependence are assumed to be

$$A_{\sigma,\boldsymbol{k}} \propto k^{\lambda} \quad \text{and} \quad \omega_{\sigma,\boldsymbol{k}} \propto k^{\mu}. \tag{5.8}$$

$\lambda = 1$ and $\mu = 0$ for Fröhlich interaction

$\lambda = 1/2$ and $\mu = 1$ for deformation potential type interaction

They showed that the stability of the free state critically depends on the index defined by

$$\eta = \delta - 2\lambda - 2. \tag{5.9}$$

The results are summarized as

when $\eta > 0$, an electron in free $\psi_{k=0}$ state is stable, an example is the short-range interaction with deformation potential ($\lambda = 0$ in $\delta = 3$-dim.)

when $\eta < 0$, an electron in free $\psi_{k=0}$ state is unstable, an example is the long-range interaction with Fröhlich interaction ($\lambda = 1$ in $\delta = 3$-dim.)

These also agree with the results in Chapter 4.

5.2 Optical Absorption in Perfect Crystals

5.2.1 Urbach–Martiessen Rule

Next, we study how the electron–lattice interaction affects the photoabsorption spectra in a perfect crystal. In the initial state of the photoabsorption, there is no carrier in the material; then the electron–lattice interaction is absent and has no effect. On the contrary, the final state of the photoabsorption, carrier(s) is created in the material; then the electron–lattice interaction is switched. Thus, the photoabsorption spectrum is affected by the electron–lattice interaction. It would be worthy to remember here the exciton effect in the photoabsorption, where the Coulomb interaction is switched between a created conduction electron and valence hole. At the lower side of the fundamental photoabsorption energy band, there appears a sharp exciton band located at E_{exciton}, which is below the band gap energy by the exciton binding energy. We will see that the optical spectra reflect the electron–lattice interaction in the final state of the optical transition.

It has been found experimentally in various nonmetallic materials that the exciton absorption spectra $I_{abs}(h\nu)$ at lower side show an exponential tail, which has a form:

$$I_{abs}(h\nu) = I_0 \exp\left[-\frac{\sigma(E_0 - h\nu)}{k_B T}\right]. \tag{5.10}$$

Here, σ is a material dependent dimensionless constant of the order of unity, called a steepness parameter, k_B the Boltzmann constant, T absolute temperature, and E_0 the converging point energy. This dependence, called the Urbach rule [3] or Urbach–Martiessen (U–M) rule [4], has been shown to found in various nonmetallic materials [5] (Fig. 5.2). A similar exponential tail in the optical absorption spectrum is also found in various amorphous semiconductors and glasses, where excitons are considered to feel statistically randomly distributed potentials in space. These facts suggest that there is the same root of this U–M rule: The exciton is trapped by randomness caused by thermal fluctuation of the lattice in crystal materials or statistical fluctuation in amorphous materials. Figure 5.3 illustrates a proposed mechanism of the U–M rule.

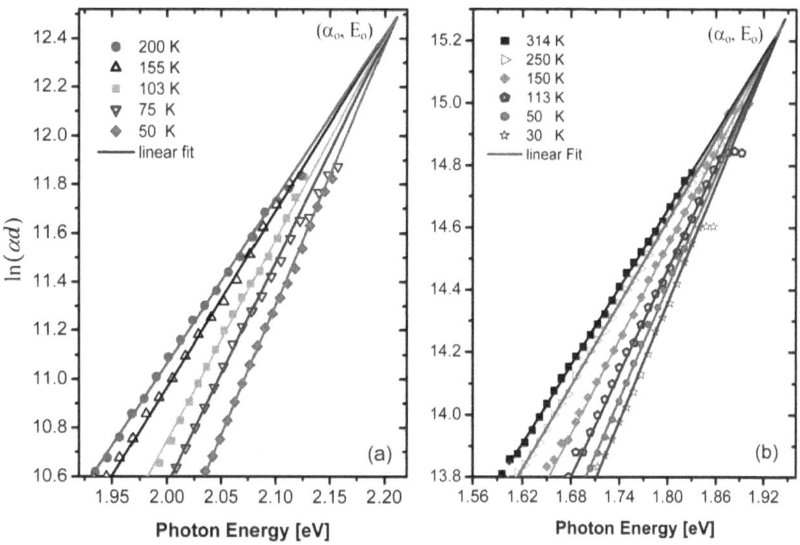

Figure 5.2 Absorption spectrum showing the Urbach–Martiessen rule near the fundamental edge of (a) ZrS_2 and (b) $ZrS_{0.5}Se_{0.5}$ single crystal. Reproduced with permission from Ref. [5].

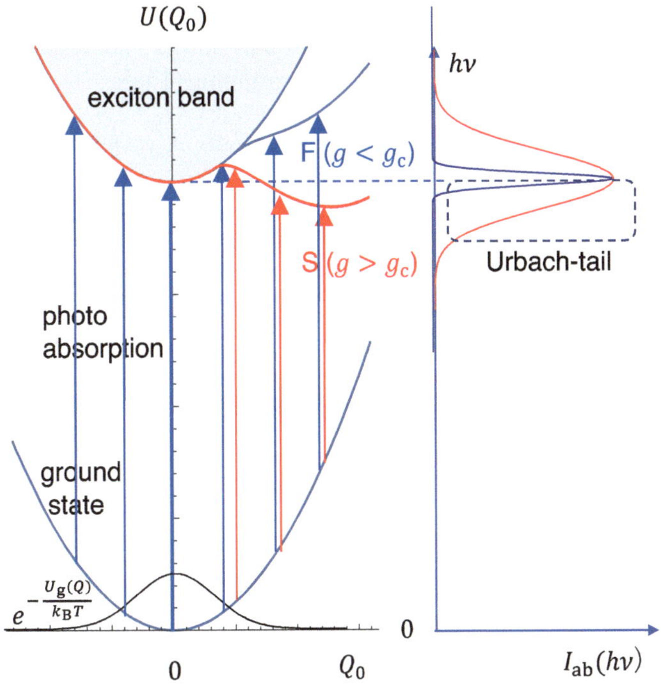

Figure 5.3 Absorption processes in the C.C. diagram: transient trapped excitons are considered to be the origin of the Urbach-Martiessen tail just below the fundamental edge.

Various theoretical investigations have been proposed [6]. Among them, Sumi [7] tried to analyze the spectrum with a model similar to Eq. (5.1). There the effect of the random potential is taken into account using dynamical CPA. Sumi successfully explained the U–M rule in the absorption spectrum; however, it involves several approximations. These defects have been removed by Schreiber [8]. He has completed to obtain exact numerical solutions of the Hamiltonian matrix using the Lanczos method in computer simulation. Figure 5.4 shows the obtained results for one, two, and three dimensions.

The obtained numerical results show us that the steepness parameter in Eq. (5.10) has a relation with the coupling constant $g \equiv E_{LR}/B$ of the exciton-lattice interaction as

$$\sigma = s/g. \tag{5.11}$$

Here, s is a numerical constant order of unity: $s = 1.5$ for the simple cubic lattice and $s = 1.24$ for the square lattice. These values are consistent with the criterion for self-trapping (Table 5.1).

Table 5.1 Relation among steepness parameter σ, coupling constant g, and the character of relaxed exciton

Material	Steepness σ	Coupling g	Relaxed exciton
alkali halides	~0.8	~1.87	self-trap
CdS	2.2	0.68	free
$AgCl_xBr_{1-x}$	1.0 → 0.8	0.66 → 0.83	free → self-trap

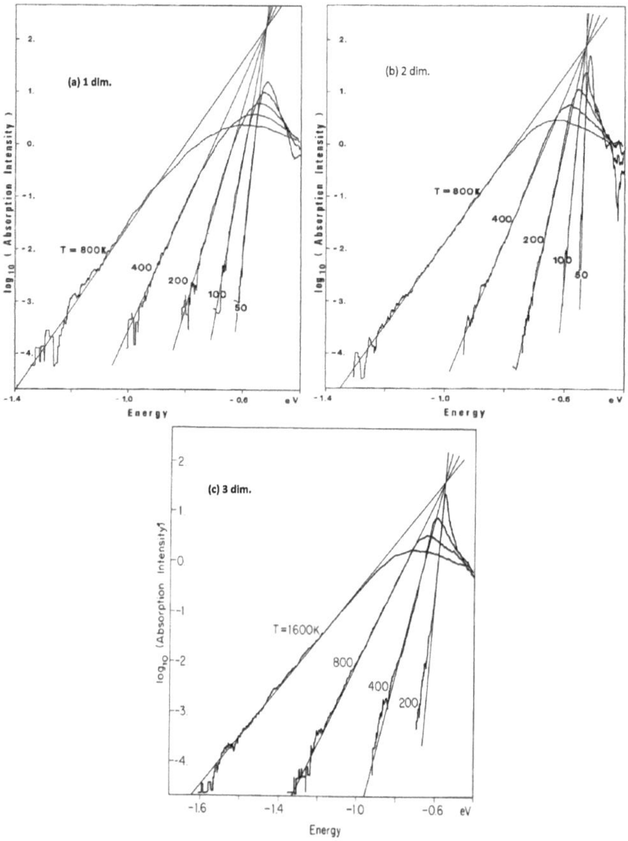

Figure 5.4 Numerically calculated absorption spectrum for (a) one-, (b) two-, and (c) three-dimensional lattice at various temperatures. After [8] with minor revision.

5.2.2 Relation with Emission Spectra

We have learned in Section 3.4.2 and Appendix II that the Einstein relation holds between the photoabsorption spectrum and photoemission spectrum.

$$I_{\mathrm{em}}(h\nu) \propto \exp\left(-\frac{h\nu}{k_{\mathrm{B}}T}\right) I_{\mathrm{ab}}(h\nu). \tag{5.12}$$

Then, if we apply the rule to the Urbach energy region in the optical absorption, we could obtain the emission spectrum as

$$I_{\mathrm{em}}(h\nu) \propto \exp\left(-\frac{h\nu}{k_{\mathrm{B}}T}\right) I_{\mathrm{ab}}(h\nu) \propto \exp\left(\frac{(\sigma-1)h\nu}{k_{\mathrm{B}}T}\right). \tag{5.13}$$

This form suggests the following conclusion:

(1) When $\sigma < 1$, the emission spectrum exhibits a decreasing function of $h\nu$; this is consistent with free exciton annihilation: F-type.

(2) When $\sigma > 1$, the emission spectrum exhibits an increasing function of $h\nu$; this is consistent with self-trapped exciton annihilation, showing a gaussian spectrum with a large Stokes shift: S-type.

These statements are consistent with experimental and also theoretical results (Table 5.1).

References

1. Toyozawa, Y., *Optical Processes in Condensed Matter* (Springer-Verlag, Berlin, 2002).
2. Toyozawa, Y., and Shinozuka,Y., *J. Phys. Soc. Jpn.*, **48** (1980) 472.
3. Urbach, F., *Phys. Rev.*, **92** (1953) 1324.
4. Martiessen, W., *J. Phys. Chem. Solids*, **2**(1957) 257.
5. Moustafa, M., Wasnick, A., Janowitz, C., and Manzke, R., *Phys. Rev. B*, **95** (2017) 245207.
6. Dow. J. D., and Redfield, D., *Phys. Rev. B*, **5**(1972) 594.
7. Sumi, H., and Toyozawa, Y., *J. Phys. Soc. Jpn.*, **31**(1971) 342.
8. Schreiber, M., and Toyozawa, Y., *J. Phys. Soc. Jpn.*, **51**(1982) 1544.

Chapter 6

Electron–Lattice Interaction at Deep-Level Defects

To investigate the dynamics of various electronic-atomic processes in condensed matter, it is convenient to introduce the configuration coordinate diagram (CCD), in which the adiabatic potential of the system is presented as a function of the atomic configurations. The CCD was first introduced for molecular systems, and later it has also been used for localized states in insulators and then to deep-level defects (impurity) in semiconductors [1, 2], and succeeded in explaining, for example, a large Stokes shift in the optical spectra, difference between the thermal and the optical depths of a trapped carrier, metastability, and photo-structural change at defects. Deep-level defects in semiconductors can capture and emit electrons and holes. The adiabatic potentials at such defects, however, cannot be given by a sum of carrier energies and the lattice potential energy, because the carrier number is not conserved in case of recombination. Also, it is uncertain for deep levels how to refer to the direction of energy as a conduction electron (upward) or a valence hole (downward).

The purpose of Chapter 6 is to show how to present the correct configuration coordinate diagrams for deep levels in relation with both the conduction and valence bands, and to

Electron–Lattice Interactions in Semiconductors
Yuzo Shinozuka

ISBN 978-981-4800-96-9 (Hardcover), 978-1-003-02346-3 (eBook)
www.jennystanford.com

describe consistently the whole recombination process of an electron and a hole via a deep level. Hereafter, in order to avoid confusion, otherwise stated, we refer to one-electron energy for electron and the increasing direction is always upward. The choice of the origin does not affect the following analysis and the results.

6.1 Deep-Localized States Couple with Conduction or Valence Band

6.1.1 One-Electron Scheme

We are familiar with the one-electron energy scheme in semiconductor physics (see Fig. 1.1(C)). In a perfect crystal, one-electron state $\psi_{\sigma,\boldsymbol{k}}(\boldsymbol{r})$ and its energy $\varepsilon_{\sigma,\boldsymbol{k}}$ are defined with a band index σ and a wave vector $\boldsymbol{k}$. If the translational symmetry of a crystal is broken by a defect (impurity, vacancy, ...), there appears a spatially localized electronic state(s) at the defect. Suppose that the defect potential gradually increases from zero. Depending on the sign of the defect potential, a localized state(s) splits off from the conduction band or valence band. In Fig. 6.1, the energy of a conduction electron is considered to increase upward and that of a valence hole increases downward. The light blue (beige) hatching represents continuum spectrum of a free electron (hole) states. The abscissa is sometimes supposed to be a spatial coordinate or location in the material. We can put several electrons and holes at a time in this figure and add some localized electrons and holes at defects. Electronic processes in semiconductors can be described with this scheme, except for those induced by strong electron–lattice interaction [3].

6.1.2 Trapping of a Conduction Electron

6.1.2.1 General

We first discuss a trapping process of a conduction electron by a point-defect and the effect of the electron–lattice interaction. In principle, these phenomena can be treated as an $N_e + 1$ electron problem in the scheme (B) in Section 1.1. The adiabatic potentials

$U_l^{N_e+1}(\{\boldsymbol{R}\})$ for the l-th state of $N_e + 1$ electrons can be calculated in principle by the following steps as a function of all the atomic displacement $\{\boldsymbol{R}\}$ including those of the defect.

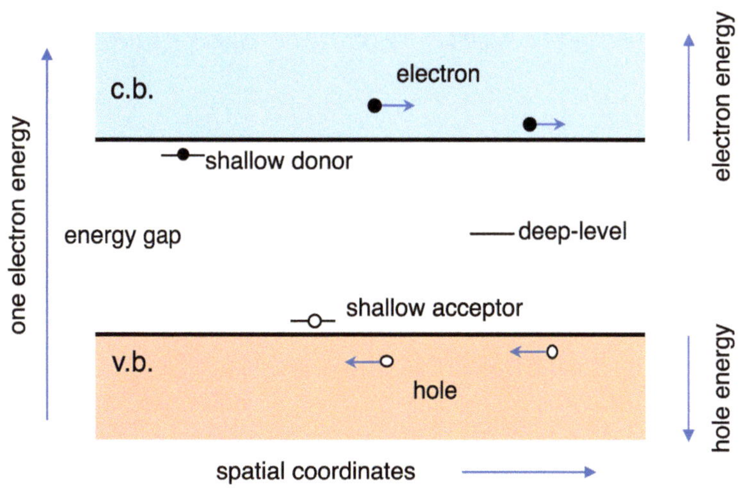

Figure 6.1 One-electron energy scheme for semiconductors (see Fig. 1.1(C)). Light blue (beige) hatching represents continuum spectrum of the conduction (valence) band. This is a familiar diagram in semiconductor physics, where we can put several electrons and holes in it at a time.

(a) N_e electrons occupy the lowest possible one-electron states and, hence, fill all the valence band and take a part in the elastic lattice. Under the harmonic approximation, $U_g^{N_e}(\{\boldsymbol{R}\})$ can be diagonalized with normal modes $\{q_k\}$ whose frequency $\{\omega_k\}$, as has seen in Chapter 1.

$$U_g^{N_e}(\{\boldsymbol{R}\}) = \text{const.}\,(= 0) + \frac{1}{2}\sum_k \omega_k^2 q_k^2. \tag{6.1}$$

Most modes $\{q_k\}$ are similar to those in the relevant crystal without defect, while some modes may be localized around the defect: The index k stands for localized mode as well as a wave vector, and also the mode index σ, if necessary.

(b) An extra electron occupies the l-th one-electron state, mainly constructed from the conduction band, with an energy $\varepsilon_l\{\boldsymbol{R}\}$ which depends on the atomic configuration $\{\boldsymbol{R}\}$. (The index

l is originally referred to as $N_e + 1$ electron states, but under the above assumption, it can stand for the one-electron state). Then, our task is to solve the Shrödinger equation

$$[\frac{\boldsymbol{p}^2}{2m} + V_{\text{eff}}(\boldsymbol{r}) + V_{\text{def}}(\boldsymbol{r}_\text{e}) + V_{\text{eL}}(\boldsymbol{r}_\text{e}, \{\boldsymbol{R}\})]\psi_{l,\{\boldsymbol{R}\}}(\boldsymbol{r}_\text{e}) = \varepsilon_l(\{\boldsymbol{R}\})\psi_{l,\{\boldsymbol{R}\}}(\boldsymbol{r}_\text{e}). \tag{6.2}$$

The first two terms in the bracket is one electron Hamiltonian $h_\text{e}^{\text{eff}} = \boldsymbol{p}^2/2m + V_{\text{eff}}(\boldsymbol{r})$ in a perfect crystal, which can be replaced by $\boldsymbol{p}^2/2m^*$ if the effective mass approximation is applicable, see Section 2.2.2. There are several suitable theoretical approaches to calculate $\varepsilon_l\{\boldsymbol{R}\}$.

Since $V_{\text{def}}(\boldsymbol{r}_\text{e})$ is spatially localized, the possible one-electron states are

l = (f.e.): characterized by the wave vector k, moves freely through the host crystal, $\{\boldsymbol{R}\}$ dependence is infinitesimal.

l = (t.e.): localized at the defect, the character depends on V_{def} and V_{eL}, strong $\{\boldsymbol{R}\}$ dependence.

Thus, the adiabatic potentials are calculated as a function of $\{\boldsymbol{R}\}$ or the normal coordinate $\{q_k\}$.

$$U_l^{N_\text{e}+1}(\{\boldsymbol{R}\}) = \varepsilon_l(\{\boldsymbol{R}\}) + U_\text{g}^{N_\text{e}}(\{\boldsymbol{R}\}) = \varepsilon_l(\{q_k\}) + \frac{1}{2}\sum_{k,\sigma}\omega_k^2 q_k^2. \tag{6.3}$$

The presence of a defect (impurity, vacancy, etc) may also affect the position of surrounding host atoms. The effect of displacements $\{\bar{\boldsymbol{R}}\}_{\text{def.g}}^{N_\text{e}} - \{\bar{\boldsymbol{R}}\}_\text{g}^{N_\text{e}}$ relative to those without defect should be treated as a static distortion, which can be included in the static defect potential $V_{\text{def}}(\boldsymbol{r}_\text{e})$.

6.1.2.2 Simple situation

A simple situation would be a neutral point defect which attracts a conduction electron with a short range e–L interaction. The tight binding model introduced in Section 4.3.3 can represent the situation, where the Hamiltonian is given by Eq. (4.29–30)

$$h_e + V_{def} + V_{eL} + U_L$$
$$= \sum_n |\phi_n^c\rangle \, \varepsilon^c \langle \phi_n^c| + \sum_n \sum_m |\phi_n^c\rangle \, t_{n,m}^c \langle \phi_m^c|$$
$$+|\phi_0^c\rangle(\varepsilon_d^c - \varepsilon^c)\langle \phi_0^c| - \sum_n |\phi_n^c\rangle \beta_c Q_n \langle \phi_n^c| + \frac{1}{2}\sum_n C Q_l^2. \quad (6.4)$$

Here ϕ_n^c is the Wannier orbital at the n-th site, whose site energy is $\varepsilon^c = 0$ except at the defect $n = 0$: $\Delta_c (\equiv \varepsilon_d^c - \varepsilon^c) < 0$ is the static potential depth of the defect. The electronic state couples with the lattice distortion Q_n by a short-range interaction with a coupling constant β_c.[1] Other $Q_{n\neq 0}$'s do not affect remarkably, then we will neglect hereafter. The electron transfer $t_{n,0}^c$ between the defect ϕ_0^c and the host ϕ_n^c could be different from the corresponding $t_{n+m,m}^c$ between the hosts.

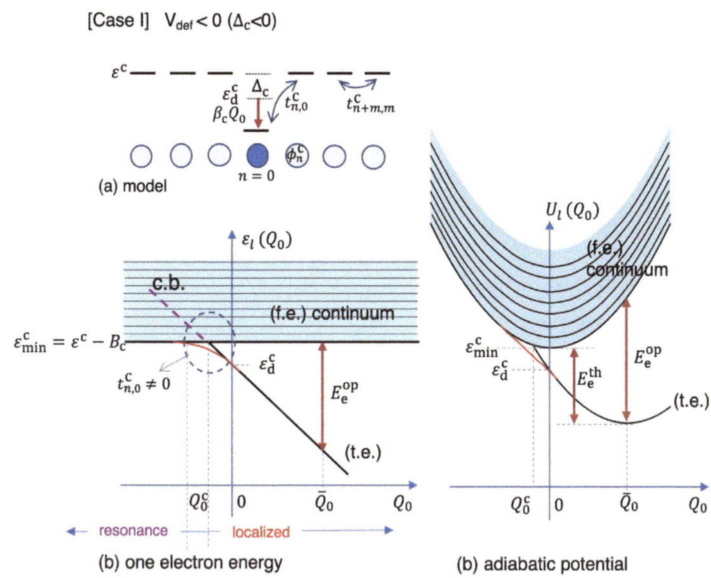

Figure 6.2 (a) a simple tight binding model for a conduction electron. (b) One-electron energy $\varepsilon_l(Q_0)$, and (c) adiabatic potentials $U_l(Q_0)$ as a function of the atomic coordinates $\{\Delta R_i\} \rightarrow Q_0$ for a neutral point defect which attracts a conduction electron ($V_{def} < 0$). The light blue hatching represents a continuous spectrum of a free electron in the conduction band showing a set of parabolas in (c) whose minimum are all located at $Q_0 = 0$.

[1]One can use "$+\beta_c Q_n$" in Eq. (6.4), since the lattice Q_n can deform both + and −.

One-electron energy $\varepsilon_l(Q_0)$ can be calculated by a procedure in Appendix IV.

$$(h_e + V_{eL})\psi_l^c(Q_0) = \varepsilon_l(Q_0)\psi_l^c(Q_0). \qquad l = 1, \ldots, N \tag{6.5}$$

The suffix l runs from an increasing order of energy. Note that a complete set of eigenstate $\{\psi_l^c(Q_0)\}$ are constructed from a set of Wannier states of the conduction band $\{\phi_n^c\}$.

When the potential strength $|\Delta_c - \beta_c Q_0|$ is large compared to B_c, a half of the conduction band width, the defect generates a localized state ψ_{loc}^c below the conduction band, hence in the band gap (Figure 6.2b). Thus, a conduction electron can occupy either

l = (f.e.): free state ψ_k^c similar to Bloch states with a wave vector k. $\varepsilon_l(Q_0) = \varepsilon_k^c$ has continuum spectrum and Q_0-dependence is infinitesimal. The lowest energy is $\varepsilon_{min}^c = \varepsilon^c - B_c$

l = (t.e.): a localized state ψ_1^c mainly constructed from ϕ_0^c. $\varepsilon_l(Q_0) \sim \varepsilon_d^c - \beta_c Q_0$, a single isolated level.[2]

Adding the elastic potential energy, the adiabatic potentials are given by

$$U_l(Q_0) = \varepsilon_l(Q_0) + \frac{1}{2}CQ_0^2. \qquad l = (\text{f.e.}) \text{ and } (\text{t.e.}) \tag{6.6}$$

Here, C is the elastic constant. $U_{f.e.}(Q_0)$ consists of a set of many parabolas, whose minima are all located at $Q_0 = 0$. The minimum of $U_{t.e.}(Q_0)$ is located at $\bar{Q}_0 = \alpha_c$ which differs that of $U_{f.e.}(Q_0)$. Then the optical depth E_e^{op} becomes $\beta_c^2/2C$ larger than the thermal depth E_e^{th}, where $\beta_c^2/2C$ (=E_{LR}) is the lattice relaxation energy.

$E_e^{op}(= \varepsilon_d^c - \varepsilon_{min}^c + \beta_c^2/C)$: the optical depth of a trapped electron defined by the energy necessary for optical ionization, $U_{f.e.}(\bar{Q}_0)$ - $U_{t.e.}(\bar{Q}_0)$ at the minimum of $U_{f.e.}(Q_0)$ (Franck-Condon principle).

[2]When the transfer $t_{n,0}^c$ between ϕ_0^c and the other ϕ_n^c's are non zero, $\varepsilon_1(Q_0)$ deviates from a straight Q_0 dependence (Δ - $\beta_c Q_0$) as indicated by a red line in Fig. 6.2(b). Accordingly, the wave function ψ_1^c extends in space and the crossing point Q_0^c shifts left. For distortion $Q < Q_0^c$ the localized one-electron state disappears, irrespective whether ϕ_0^c at the defect is originated from the host Wannier orbitals or impurity extra orbital.

$E_e^{th}(= \varepsilon_d^c - \varepsilon_{min}^c + \beta_c^2/2C)$: the thermal depth of a trapped electron defined by the energy necessary for thermal ionization. It is also appeared in the Boltzmann factor of the electron population ratio at the defect $n_{unocc}/n_{occ} = \exp\left[-E_e^{th}/k_B T\right]$ at temperature T.

If the lattice Q_0 distorts to $Q_0^c = (\varepsilon_d^c - \varepsilon_{min}^c)/\beta_c$ where the two adiabatic potentials cross each other, an electron transition is possible between (f.e.) and (t.e.) without photon emission/absorption. This nonradiative process will be discussed thoroughly in Chapter 7.

6.1.3 Localized Electron Deep into VB: Incorrect CCD

Consider the lattice relaxation energy is large compared to the band gap energy. After a conduction electron is captured, its level $\varepsilon_{t.e.}(Q_0)$ decreases in the band gap and may reach the top of valence band. Then the trapped electron has a chance to recombine with a valence hole nonradiatively. In order to discuss this process, one is likely to calculate adiabatic potentials similar to Eqs. (6.3) or (6.6),

$$U_l(Q_0) = \varepsilon_l(Q_0) + \frac{1}{2}CQ_0^2, \tag{6.7}$$

extending one-electron energy $\varepsilon_l(Q_0)$ covering both the conduction and valence bands (Fig. 6.3), and focus following three states.

(I) conduction band minimum: $\varepsilon_{CBM}(Q_0) \rightarrow$ free electron (+ free hole)

(II) localized state: $\varepsilon_{t.e.}(Q_0) \rightarrow$ trapped electron (+free hole)

(III) valence band maximum: $\varepsilon_{VBM}(Q_0) \rightarrow$ no carrier

These types of adiabatic potentials (Fig. 6.3b) are frequently found in literature, including a pioneering paper by Henry and Lang [4]. There, a transition from (I) to (II) is assigned as an electron capture and that from (II) to (III) is assigned as a hole capture resulting in recombination. However, Eq. (6.7) and Fig. 6.3b are confusing, for example:

(a) In Fig. 6.3b, where is a hole in the (II) state? How to include the hole freedom? The diagram lacks the electron–hole symmetry.

(b) Is the direction of a hole energy opposite to that of an electron? In which direction does the lattice relax after a hole is captured?

(c) Can we put several electrons and holes at a time in Fig. 6.3b as shown in Fig. 6.1?

(d) Since there is a finite mixing between a localized state ϕ_0^c and valence band states, i.e., transfer energy t_{n0}^{vc} is nonzero between ψ_1^c and Wannier states $\{\phi_n^v\}$, this makes $\varepsilon_{t.e.}(Q_0)$ pushed up from the top of the valence band and extend the wave function ψ_1^c, finally localization disappears for $Q_0 > Q_0^v$. Accordingly, the minimum $\bar{Q}_0$ of $U_{t.e.}(Q_0)$ never reaches Q_0^v, i.e., the relaxed occupied localized state remains in the band gap for any strong electron–lattice interactions (Fig. 6.3c).

These questions will be solved in Section 6.2.2.

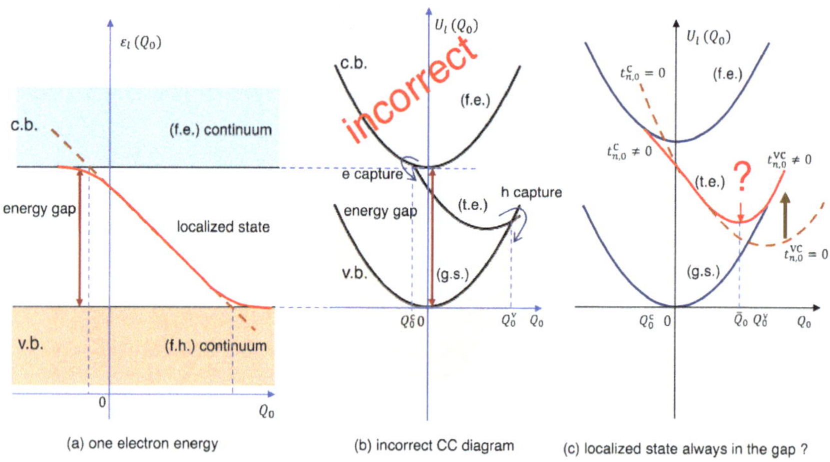

Figure 6.3 (a) one electron energy $\varepsilon_l(Q_0)$ as a function of the lattice distortion Q_0 for a deep-level defect in semiconductors. (b) Incorrect CCD, where $-\varepsilon_l(Q_0)$ in the valence band is supposed to be the energy of a hole. (c) When the interband mixing $t_{n0}^{vc} \neq 0$, does it push the localized level above the top of the valence band?

Then, how do we calculate the adiabatic potentials taking care of the electron–hole symmetry? One may try an energy sum of an electron in the conduction band, a hole in the valence band and the elastic lattice.

$$U_{ll'}(\{R_j\}) = \varepsilon_e(\{R_j\}') + \varepsilon_h(\{R_j\}') + \frac{1}{2}\sum_k \omega_k^2 q_k^2, \tag{6.8}$$

where $\varepsilon_e(\{R_j\}) = \varepsilon_l(\{R_j\})$ and $\varepsilon_h(\{R_j\}) = -\varepsilon_{l'}(\{R_j\})$. However, this does not work because the number of electrons and holes is not conserved after recombination, then it slips out the Scheme (6.8). In addition, it is unclear whether the occupation of the localized state in the band gap should be referred to an electron or a hole.

So, we should investigate the trapping of a valence hole very carefully.

6.1.4 Trapping of a Valence Hole

6.1.4.1 General

If the defect potential is repulsive for electrons, a localized state(s) splits off from the valence band. As discussed in Chapter 2, we can treat the system as $N_e - 1$ electrons. The phenomena "trapping of a valence hole" can be examined in principle by calculating adiabatic potentials $U_{l'}^{N_e-1}(\{\boldsymbol{R}\})$ for l'-th state of $N_e - 1$ electrons as a function of the atomic configuration $\{\boldsymbol{R}\}$ at the defect.

(a) $N_e - 1$ electrons, respectively, occupy the lowest possible one-electron states in the valence band.

(b) Only the l'-th one-electron state is empty whose one-electron energy is $\varepsilon_{l'}(\{\boldsymbol{R}\})$. If we further add an electron into the l'-th one-electron state which needs $\varepsilon_{l'}(\{\boldsymbol{R}\})$; this gives a complete filled valence band. Then we have a relation

$$U_g^{N_e}(\{\boldsymbol{R}\}) = U_{l'}^{N_e-1}(\{\boldsymbol{R}\}) + \varepsilon_{l'}(\{\boldsymbol{R}\}). \tag{6.9}$$

Thus, the adiabatic potentials are calculated as a function of $\{\boldsymbol{R}\}$ or $\{q_k\}$.

$$U_{l'}^{N_e-1}(\{\boldsymbol{R}\}) = -\varepsilon_{l'}(\{\boldsymbol{R}\}) + U_g^{N_e}(\{\boldsymbol{R}\}) = -\varepsilon_{l'}(\{q_k\}) + \frac{1}{2}\sum_k \omega_k^2 q_k^2. \tag{6.10}$$

If the effective mass approximation is applicable, we can use Eq. (6.2) with replacing h from e. One can discuss a hole trapping process similar to Section 6.1.2. However, we proceed carefully by reviewing the concept "hole" once again.

6.1.4.2 Simple situation

A simple situation would be a neutral point defect which repulses valence electrons with a short-range e–L interaction. The following tight binding model (Fig. 6.4a) can represent the situation:

$$h_{\mathrm{h}} + V_{\mathrm{eh}} = \sum_{n} |\phi_n^{\mathrm{v}}\rangle \varepsilon_n^{\mathrm{v}} \langle \phi_n^{\mathrm{v}}| + \sum_{n}\sum_{m} |\phi_n^{\mathrm{v}}\rangle\, t_{n,m}^{\mathrm{v}} \langle \phi_m^{\mathrm{v}}|. \tag{6.11}$$

Here the site energy has a form

$$\varepsilon_n^{\mathrm{v}} = \begin{cases} \varepsilon^{\mathrm{v}} + \Delta_{\mathrm{v}} - \beta_{\mathrm{v}} Q_0 & n = 0\,^{3} \\ \varepsilon^{\mathrm{v}} & n \neq 0 \end{cases}, \tag{6.12}$$

This Hamiltonian is similar to Eq. (6.4), but ϕ_n^{v} represents a Wannier orbital of the valence band. An electron feels the potential height $\Delta_{\mathrm{v}} > 0$ of the defect at the origin and interacts with Q_0 the lattice distortion with a coupling constant β_{v}. Modes $Q_{n\neq 0}$ at the other sites are neglected. The electron transfer $t_{n,0}^{\mathrm{v}}$ between the defect ϕ_0^{v} and the host ϕ_n^{v} would be different from the corresponding $t_{n+m,m}^{\mathrm{v}}$ among the hosts. One-electron eigenenergy $\varepsilon_l^{\mathrm{v}}$ can be obtained by the same procedure discussed in the previous section. The suffix l runs from an increasing order of energy.

When the potential strength $|\Delta_{\mathrm{v}} - \beta_{\mathrm{v}} Q_0|$ is large compared to B_{v}, a half of the valence band width, the defect generates a localized state ψ_N^{v} above the valence band, hence in the band gap (Fig. 6.4b). Then, the possible eigenstates for a hole are

ψ_k^{v}: free states similar to bulk Bloch states with a wave vector k.

ψ_N^{v}: a localized state, mainly constructed from $\sim\phi_0^{\mathrm{v}}$.

Note that a complete set of $\{\psi_k^{\mathrm{v}}\}$ and $\psi_{\mathrm{loc}}^{\mathrm{v}}$ is constructed from a set of Wannier states of the valence band $\{\phi_n^{\mathrm{v}}\}$. Then, when ψ_N^{v} is occupied by an electron, this does not mean any net charge localization around the defect. Remember that a complete set of ψ_k^{v} and $\psi_{\mathrm{loc}}^{\mathrm{v}}$'s is constructed from a complete set of Wannier states of the valence band $\{\phi_l^{\mathrm{v}}\}$. When ψ_N^{v} is occupied, the

[3] One can use "$-\beta_{\mathrm{v}} Q_0$" in Eq. (6.12), since the lattice Q_0 can deform both + and –.

N_e-electron state corresponds to the completely filled valence band, resulting a uniform charge distribution of valence electrons (see Section 1.5.5). Then the physically meaningful description should be that when ψ_N^v is empty, a hole is localized at the defect with positive charge spatial distribution given by $|\psi_N^v|^2$.

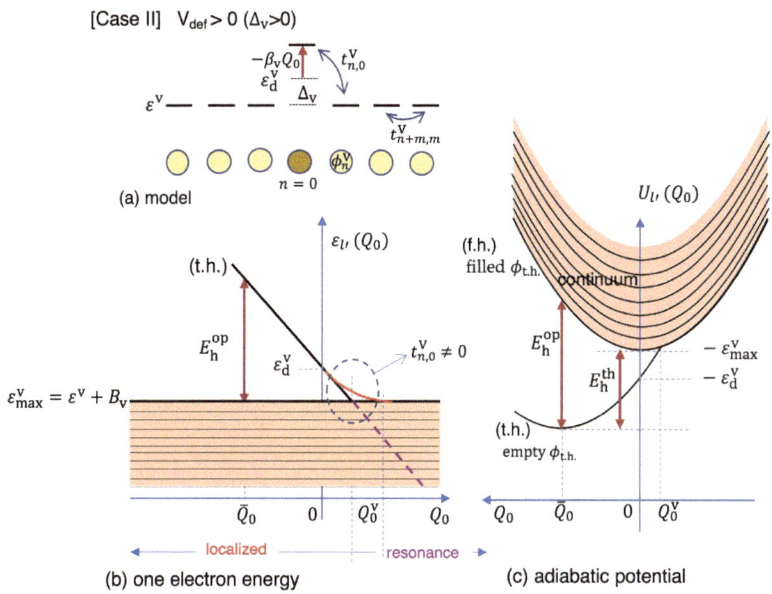

Figure 6.4 (a) A simple tight binding model for valence electron. (b) One-electron energy $\varepsilon_l(Q_0)$ and (c) adiabatic potentials $U_{l'}(Q_0)$ as a function of the interaction mode Q_0 for a neutral point defect which attracts a valence hole ($V_{def} > 0$). The light beige hatching represents a continuous spectrum of a free hole in the valence band showing a set of parabolas in (c) whose minimums are all located at $Q_0 = 0$.

Thus, in the hole picture, a valence hole can occupy either

l' = (f.h.): a free state: $-\varepsilon_{l'}(Q_0) = -\varepsilon^v_{k'}$, Q_0 dependence is infinitesimal, continuum spectrum.

l' = (t.h.): a localized state: $-\varepsilon_{l'}(Q_0) \sim -\Delta_v + \beta_v Q_0$, an isolated level[4]

---

[4]When the transfer $t^v_{n,0}$ between ϕ^v_0 and the other $\{\phi^v_n\}$ is nonzero, $\varepsilon_l(Q_0)$ deviates from a straight Q_0 dependence ($\sim\Delta_v + \beta_v Q_0$) as indicated by a red line in Fig. 6.4. Accordingly, the crossing point Q^v_0 shifts right. For $Q < Q^v_0 <$ (<0) configuration, ϕ^v_0 resonances with host valence band and all eigenstates are extended, irrespective whether ϕ^v_0 at the defect is composed of the host Wannier orbitals or impurity extra orbital.

With the obtained eigenenergy $-\varepsilon_{l'}(Q_0)$ of a hole, adding an elastic potential energy, the adiabatic potential is given by

$$U_{l'}(Q_0) = -\varepsilon_{l'}(Q_0) + \frac{1}{2}CQ_0^2. \qquad l' = (\text{f. h.}) \text{ and } (\text{t. h.}) \quad (6.13)$$

Figure 6.4(b) shows an example of $\varepsilon_{l'}(Q_0)$ and Fig. 6.4(c) shows the corresponding adiabatic potentials. It should be noted here that the adiabatic potentials for (f.h.) have continuum spectrum due to the kinetic energy of a free hole.

As a result of the electron–lattice interaction at a defect, the minimum of $U_{\text{t.h.}}(Q_0)$, differs from that of $U_{\text{f.h.}}(Q_0)$, and then the optical depth (ionization energy) E_{h}^{op} becomes larger than the thermal depth E_{h}^{th}.

$E_{\text{h}}^{\text{op}}(= \varepsilon_{\text{d}}^{\text{v}} - \varepsilon_{\text{max}}^{\text{v}} + \beta_{\text{v}}^2/C)$: the optical depth: energy necessary for an optically excitation of a localized hole at the defect, $U_{\text{f.h.}}(\bar{Q}_0) - U_{\text{t.h.}}(\bar{Q}_0)$ at the minimum of $U_{\text{t.h.}}(Q_0)$ (Franck-Condon principle).

E_{h}^{th} $(= \varepsilon_{\text{d}}^{\text{v}} - \varepsilon_{\text{max}}^{\text{v}} + \beta_{\text{v}}^2/2C)$: the thermal depth: energy necessary for thermal ionization of hole-occupied defect. It is also appeared in the Boltzmann factor in the hole population ratio at the defect $n_{\text{unocc}}/n_{\text{occ}} = \exp[-E_{\text{h}}^{\text{th}}/E_{\text{h}}^{\text{op}}]$ at temperature T.

If the lattice Q_0 distorts to Q_0^{v}, where two adiabatic potentials cross each other, a nonradiative transition is possible. This hole capture process, (f.h.) → (t.h.), will be discussed thoroughly in Section 6.2.2.

6.2 Deep-Localized States Couple with Valence and Conduction Bands

6.2.1 Calculation of Localized State in the Band Gap

Now is the time to investigate the adiabatic potentials when the lattice relaxation energy is large compared to the band gap, where deep levels couple with both the conduction and valence band. At least three types of electronic states in the N_{e}-electron scheme must be considered [5, 6].

(a) The ground state (g.s.): All N_e electrons occupy the valence band $\{\psi_l^{\mathrm{v}}\}$. There is no free carrier. The defect is neutral.

(b) One free electron + one free hole (f.e.+f.h.)

(c) One trapped electron + one free hole (t.e.+f.h.) for ($V_{\mathrm{der}}(r) < 0$)

One free electron + one trapped hole (f.e.+t.h.) for ($V_{\mathrm{der}}(r) > 0$)

It is difficult to combine simply two effective mass approximations for a conduction electron and a valence hole. Then we will use a combination of two tight-binding-model, Eqs. (6.4) and (6.11) (Figs. 6.5a,b).

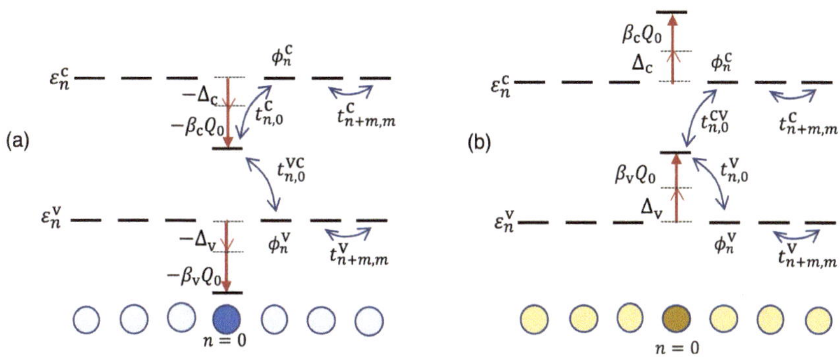

Figure 6.5 A tight binding model, where electron states are mixed via a defect with potential difference and the modulation in transfer energies. (a) A localized state generated from the conduction band (b) from the valence band.

$$H_e + V_{\mathrm{eL}} = \sum_n |\phi_n^{\mathrm{c}}\rangle \varepsilon_n^{\mathrm{c}} \langle \phi_n^{\mathrm{c}}| + \sum_n \sum_m |\phi_n^{\mathrm{c}}\rangle\, t_{n,m}^{\mathrm{c}} \langle \phi_m^{\mathrm{c}}|$$
$$+ \sum_n |\phi_n^{\mathrm{v}}\rangle \varepsilon_n^{\mathrm{v}} \langle \phi_n^{\mathrm{v}}| + \sum_n \sum_m |\phi_n^{\mathrm{v}}\rangle\, t_{n,m}^{\mathrm{v}} \langle \phi_m^{\mathrm{v}}|$$
$$+ \sum_{n\neq 0} \left(|\phi_n^{\mathrm{c}}\rangle t_{n,0}^{\mathrm{cv}} \langle \phi_0^{\mathrm{v}}| + |\phi_0^{\mathrm{v}}\rangle t_{0,n}^{\mathrm{vc}} \langle \phi_n^{\mathrm{c}}| + |\phi_n^{\mathrm{v}}\rangle t_{n,0}^{\mathrm{vc}} \langle \phi_0^{\mathrm{c}}| + |\phi_0^{\mathrm{c}}\rangle t_{0,n}^{\mathrm{cv}} \langle \phi_n^{\mathrm{v}}| \right) \tag{6.14}$$

The site energy has a form

$$\varepsilon_n^{\mathrm{c}} = \begin{cases} \varepsilon^{\mathrm{c}} + \Delta_{\mathrm{c}} - \beta_{\mathrm{c}} Q_0 & n = 0 \\ \varepsilon^{\mathrm{c}} & n \neq 0 \end{cases}, \quad \varepsilon_n^{\mathrm{v}} = \begin{cases} \varepsilon^{\mathrm{v}} + \Delta_{\mathrm{v}} - \beta_{\mathrm{v}} Q_0 & n = 0 \\ \varepsilon^{\mathrm{v}} & n \neq 0. \end{cases} \tag{6.15}$$

Here, ϕ_n^σ is a Wannier state of σ-band at the n-th site (σ = c,v). The total site number N is also the number of electrons. At the defect (n = 0), an electron in the conduction (valence) band feels the potential strength Δ_c (Δ_v), and interacts with the lattice distortion Q_0 with the coupling constant β_c (β_v). If the defect is absent, the host Wannier orbitals generate a conduction band and a valence band whose one-electron energy is

$$\varepsilon_k^\sigma = \varepsilon^\sigma + t_k^\sigma, \quad \psi_k^\sigma = \frac{1}{N}\sum_m \exp[i\boldsymbol{k}\boldsymbol{R}_n]\,\phi_n^\sigma \qquad \sigma = \mathrm{c, v} \qquad (6.16)$$

The band gap energy is

$$E_\mathrm{g} = \mathrm{Min}\,[\varepsilon_k^\mathrm{c}] - \mathrm{Max}\,[\varepsilon_k^\mathrm{v}] = \varepsilon^\mathrm{c} - B_\mathrm{c} - (\varepsilon^\mathrm{v} + B_\mathrm{v}). \qquad (6.17)$$

Here B_σ is half of the σ band width.

The eigenenergy of Hamiltonian (6.14) is obtained by solving

$$(H_\mathrm{e} + V_\mathrm{eL})\psi_l^\sigma(Q_0) = \varepsilon_l^\sigma(Q_0)\psi_l^\sigma(Q_0), \quad \sigma = \mathrm{c, v}, \quad l = 1, \dots, N. \quad (6.18)$$

The suffix l runs from an increasing order of energy both for σ = c, v. Detailed procedure to calculate one-electron energy $\varepsilon_l^\sigma(Q_0)$ as a function of the lattice distortion Q_0 is given in Appendix VII. Figure 6.6 schematically shows a typical Q_0 dependence of $\varepsilon_l^\sigma(Q_0)$.

Two notices are added here: The first is, as long as the interband mixing is absent, a complete set of eigenstate $\{\psi_l^\sigma(Q_0)\}$ are constructed from a set of Wannier states $\{\phi_n^\sigma\}$ of each σ (= c, v) band. The second is a localized state appearing in the gap originates from either the valence band (ψ_N^v: Δ_c, Δ_v > 0) or the conduction band (ψ_1^c: Δ_c, Δ_v < 0). In the former case (Δ_c, Δ_v > 0), another localized state ψ_N^v may appear above the conduction band, mainly constructed from ϕ_0^c; however, it does not affect processes in the gap. In the latter case (Δ_c, Δ_v < 0), another localized state ψ_1^v may appear below the valence band, mainly constructed from ϕ_0^v. As long as this localized state is occupied by a valence electron, the total N electron state is nothing but the completely filled valence Wannier state—then no net charge localization. Thus, we need not take care of these localizations.

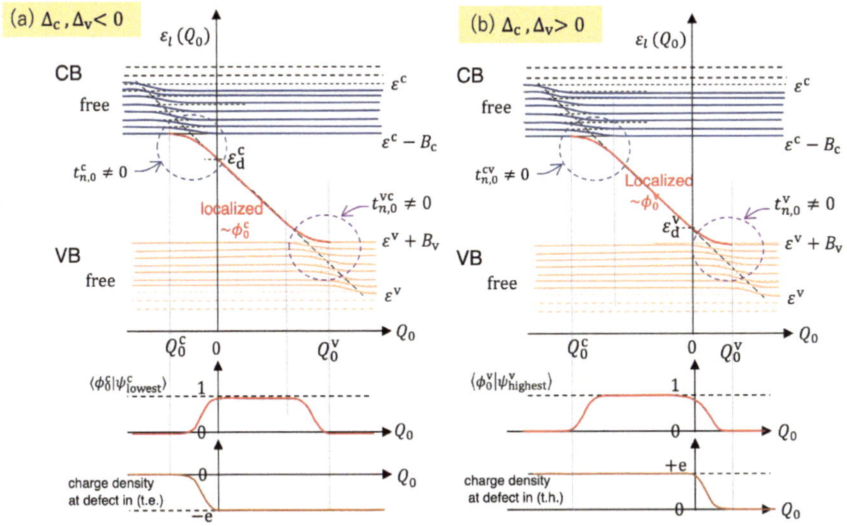

Figure 6.6 Illustration of one electron energy $\varepsilon_l^{\sigma}(Q_0)$ as a function of the lattice distortion Q_0 for a deep-level defect in semiconductors. (a) A localized state is generated from the conduction band (b) from the valence band. The transfer between the defect ϕ_0^c (ϕ_0^v) and the host bands, i.e., $t_{n,0}^{vc}(t_{n,0}^{v})\neq 0$ and $t_{n,0}^{cv}(t_{n,0}^{vc})\neq 0$, mix these states and one electron energy $\varepsilon_N^{v}(Q_0)$ ($\varepsilon_1^{c}(Q_0)$) shifts upward (downward) form the top (bottom) of the valence (conduction) band.

6.2.2 Adiabatic Potentials

A useful (but not exact) procedure to obtain the adiabatic potentials in N-electron systems would be summing up all one-electron energies.

$$U(Q_0)=\sum_{l=\mathrm{occ}}\varepsilon_l(Q_0)+V_{\mathrm{ion}}(Q_0). \tag{6.19}$$

Here the summation runs over all occupied one-electron states and V_{ion} is the interaction among ions and other miscellaneous interactions. According to the discussion in Chapter 1, $U(Q_0)$ of the ground state is nothing but the lattice elastic potential, then

$$U_{\mathrm{g.s.}}(Q_0)=\sum_{l=\mathrm{occ}}\varepsilon_l^{v}(Q_0)+V_{\mathrm{ion}}(Q_0)=\frac{1}{2}CQ_0^2. \tag{6.20}$$

with a suitable choice of the origin of the energy.

For an excited state where l' = q-th state of the valence band is empty and l = p-th state in the conduction band is occupied, the adiabatic potential is given by [2]

$$\begin{aligned} U_{\mathrm{ex}}(Q_0) &= \sum_{l \neq q} \varepsilon_l^{\mathrm{v}}(Q_0) + \varepsilon_p^{\mathrm{c}}(Q_0) + V_{\mathrm{ion}}(Q_0) \\ &= \sum_{l} \varepsilon_l^{\mathrm{v}}(Q_0) - \varepsilon_q^{\mathrm{v}}(Q_0) + \varepsilon_p^{\mathrm{c}}(Q_0) + V_{\mathrm{ion}}(Q_0) \\ &= \varepsilon_p^{\mathrm{c}}(Q_0) - \varepsilon_q^{\mathrm{v}}(Q_0) + \frac{1}{2} C Q_0^2. \end{aligned} \tag{6.21}$$

Here we neglect the Coulomb interaction between an electron and a hole for simplicity. These formulas, Eqs. (6.20) and (6.21), work well to some extent, but they are invalid when a localized state originated from the conduction (valence) band is mixed with those of the valence (conduction) band [6]. Details will be discussed in next section.

Figure 6.7 shows typical adiabatic potentials for a deep-level defect.

(a) Ground state (g.s.): The defect is neutral.

$$U_{\mathrm{g.s.}}(Q_0) = \frac{1}{2} C Q_0^2. \tag{6.22}$$

(b) One free electron + one free hole state (f.e.+f.h.):

a free state ψ_k^{c} in the conduction band is occupied $\varepsilon_p^{\mathrm{c}}(Q_0) \sim \varepsilon_k^{\mathrm{c}}$.

$N_{\mathrm{e}} - 1$ electrons occupy the valence band except $\psi_{k''}^{\mathrm{v}}$ i.e., one free hole is in the valence band, $-\varepsilon_q(Q_0) \sim -\varepsilon_{k'}^{\mathrm{v}}$. The defect is neutral.

$$\begin{aligned} U_{\mathrm{f.e.+f.h.}}(Q_0) &= \varepsilon_k^{\mathrm{c}} - \varepsilon_{k'}^{\mathrm{v}} + \frac{1}{2} C Q_0^2 \\ &= \Delta E_{\mathrm{e}} + \Delta E_{\mathrm{h}} + E_{\mathrm{g}} + \frac{1}{2} C Q_0^2. \end{aligned} \tag{6.23}$$

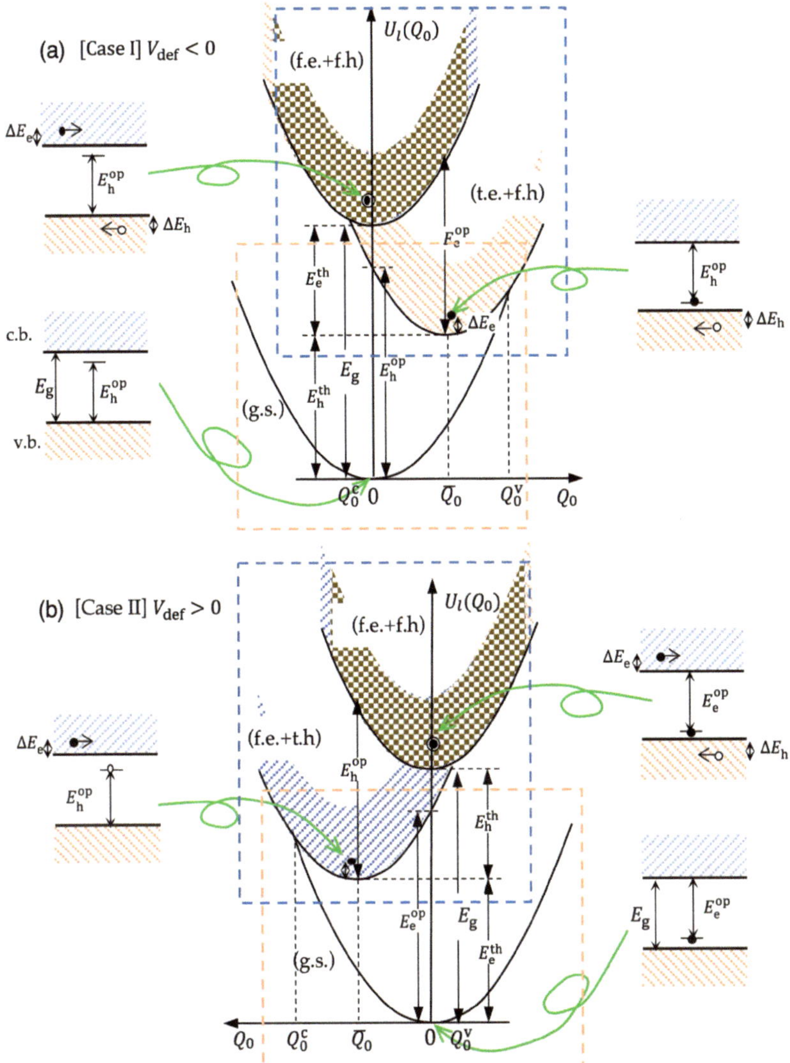

Figure 6.7 An example of the correct CCD for a point defect ((a) $V_{def}(r) < 0$ and (b) $V_{def}(r) > 0$). The light blue (beige) hatching represents a continuous spectrum of a free electron (hole), instead of parabolas shown in Figs. 6.2 and 6.4. The inserts present, respectively, the electronic configurations of (g.s.), (f.e.+f.h.) and (t.e.+f.h) or (f.e.+t.h.). Note that energy scale is reduced. Remember in (b) that full electron occupation of a localized state originated from the valence band corresponds to the uniform net charge distribution. Reproduced with permission from Ref. [3].

This has a double-continuum spectrum due to the free motions of carriers, $\Delta E_{\mathrm{e}} = \varepsilon_k^{\mathrm{c}} - \mathrm{Min}\,[\varepsilon_k^{\mathrm{c}}](> 0)$ is an electron kinetic energy and $\Delta E_{\mathrm{h}} = \mathrm{Max}\,[\varepsilon_k^{\mathrm{v}}] - \varepsilon_{k'}^{\mathrm{v}}(> 0)$ a hole kinetic energy. The lowest energy is E_{g} larger than that of $U_{\mathrm{g.s.}}(Q_0)$.

(c-1), Δ_{c}, $\Delta_{\mathrm{v}} < 0$: One trapped electron + one free hole state (t.e.+f.h.):

The defect is occupied by an electron, $\varepsilon_1^{\mathrm{c}}(Q_0)$, and charged by $-e$. $N_{\mathrm{e}} - 1$ electrons occupy the valence band except $\psi_{k''}^{\mathrm{v}}$, i.e., one free hole in the valence band. This has a single-continuum spectrum due to the free motion of a hole.

$$U_{\mathrm{t.e.+f.h.}}(Q_0) = \varepsilon_1^{\mathrm{c}}(Q_0) - \varepsilon_{k'}^{\mathrm{v}} + \frac{1}{2}CQ_0^2$$

$$= \varepsilon_1^{\mathrm{c}}(Q_0) - \mathrm{Max}\,[\varepsilon_k^{\mathrm{v}}] + \Delta E_{\mathrm{h}} + \frac{1}{2}CQ_0^2. \qquad (6.24)$$

(c-2) Δ_{c}, $\Delta_{\mathrm{v}} > 0$: One free electron + one trapped hole state (f.e.+t.h.)

The defect is unoccupied, hence occupied by a hole, $-\varepsilon_N^{\mathrm{v}}(Q_0)$, and charged by $+e$. $N_{\mathrm{e}} - 1$ electrons occupy the valence band except ψ_N^{v}. This has a single-continuum spectrum due to the free motion of an electron which occupies ψ_k^{c} in the conduction band: $\varepsilon_p^{\mathrm{c}}(Q_0) \sim \varepsilon_k^{\mathrm{c}}$.

$$U_{\mathrm{f.e.+t.h.}}(Q_0) = \varepsilon_k^{\mathrm{c}} - \varepsilon_N^{\mathrm{v}}(Q_0) + \frac{1}{2}CQ_0^2$$

$$= \Delta E_{\mathrm{e}} + \mathrm{Min}\,[\varepsilon_k^{\mathrm{c}}] - \varepsilon_N^{\mathrm{v}}(Q_0) + \frac{1}{2}CQ_0^2. \qquad (6.25)$$

At a first glance, Fig. 6.7a seems to be almost the same as Fig. 6.3b, except for the continuum spectrum of a free hole. However, in the correct diagram, each point stands for the total electronic state for a fixed configuration Q_0 or $\{R\}$, as is shown in the inserts in Fig. 6.7. It does not merely represent one carrier (an electron or a hole) as in Fig. 6.3a.

The difficulty in understanding the CC diagram for deep-level defects would be related to the interband mixing.

Case I (Δ_{c}, $\Delta_{\mathrm{v}} < 0$): crossing of (t.e.+f.h) and (g.s.)

After the first capture of a conduction electron into a localized state constructed from CB, i.e., $\psi_1^{\mathrm{c}} \sim \phi_0^{\mathrm{c}}$, the lattice Q_0 distorts and its level lowers in the gap, and then reaches the VB at Q_0^{v} (Fig. 6.6a). Then the defect state $\sim\phi_0^{\mathrm{c}}$ is mixed with those of the host VB, and the localized state ψ_1^{c} re-extends and finally merges into the body of VB: disappearance of wave function localization. Accordingly, one may consider that a trapped electron is delocalized again. But this is *not* so.

[charge localization]

Around the crossing region Q_0^{v}, all occupied states are constructed from a set of ϕ_0^{c} from CB and Wannier states $\{\phi_n^{\mathrm{v}}\}$ from VB, whose number is N. Then, according to a similar discussion on many electron state in Section 1.5.5, as long as these all N states (ψ_1^{c} + $\{\psi_k^{\mathrm{v}}\}$ beside $\psi_{k'}^{\mathrm{v}}$) are filled with N electrons, regardless each state is localized or extended, the total N electron-state is nothing but completely filled (ϕ_0^{c} + $\{\phi_n^{\mathrm{v}}\}$) + empty $\psi_{k'}^{\mathrm{v}}$; then the net electron charge remains localized at the defect (Fig. 6.6a).

[energy]

The interband mixing $t_{n,0}^{\mathrm{vc}} \neq 0$ makes $\varepsilon_1^{\mathrm{c}}(Q_0)$ up. On the other hand, the same amount of energy divided by N is equally lowered in $\varepsilon_l^{\mathrm{v}}(Q_0)$. Then

$$U_{\text{t.e.f.h.}}(Q_0) = \sum_{l \neq q} \varepsilon_l^{\mathrm{v}}(Q_0) + \varepsilon_1(Q_0) + V_{\text{ion}}(Q_0) \tag{6.26}$$

remains unchanged irrespective of whether $t_{n,0}^{\mathrm{vc}} \neq 0$ or $t_{n,0}^{\mathrm{vc}} = 0$. See additional note in Appendix V. On the other hand,

$$U_{\text{g.s.}}(Q_0) = \sum_{l=\text{occ}} \varepsilon_l^{\mathrm{v}}(Q_0) + V_{\text{ion}}(Q_0)$$

$$= \varepsilon^{\mathrm{c}} + \Delta_{\mathrm{c}} - \beta_{\mathrm{c}} Q_0 - \varepsilon_1^{\mathrm{c}}(Q_0) + \frac{1}{2} C Q_0^2, \tag{6.27}$$

decreases as indicated in Figs. 6.6 and 6.7. As long as the system moves along $U_{\text{t.e.+f.h.}}(Q_0)$, nothing happens at Q_0^{v}. During backward $Q_0(t) \to 0$, a transition from $U_{\text{t.e.+f.h.}}(Q_0)$ to $U_{\text{g.s.}}(Q_0)$ is possible at Q_0^{v}, this is nothing but a hole transition from $\psi_{k'}^{\mathrm{v}}$ to ψ_1^{c}, i.e., nonradiative capture of a free hole of the defect: a recombination has completed.

Case II: (Δ_c, $\Delta_v > 0$) crossing of (f.e.+t.h) and (g.s.)

The defect state in the gap is constructed from VB, i.e., $\psi_N^v \sim \phi_0^v$. After a hole is captured, as the lattice Q_0 distorts, the state increases its level in the gap and reaches the CB at Q_0^c (Fig. 6.6b). Then the defect state $\sim\phi_0^v$ is mixed with those of CB, and the wave function ψ_N^v of the localized state re-extends and finally merges into CB: i.e., disappearance of the wave function localization. One may consider that a trapped hole is delocalized again. But this is *not* so. Because, nevertheless, the one-electron state ψ_N^v is extended, this state is empty, which does not affect charge localization. Around the crossing region Q_0^c, the rest $N - 1$ states $\{\psi_l^v\}$ are constructed from the Wannier state $\{\phi_n^v\}$ in VB except ϕ_0^v. Then so far as these $N - 1$ states $\{\psi_l^v\}$ are filled with $N - 1$ electrons, the positive (hole) charge remains localized at the defect in proportion to $|\phi_0^v|^2$ (Fig. 6.6b). In addition, due to the interband mixing $t_{n,0}^{cv} \neq 0$, $\varepsilon_N^v(Q_0)$ is pushed down, but since the state is unoccupied, then there is no effect on $U_{\text{f.e.+t.h.}}(Q_0)$, around Q_0^c. On the contrary, for the ground state, $\varepsilon_N^v(Q_0)$ is decreased due to $t_{n,0}^{cv} \neq 0$. Then $U_{\text{f.e.+t.h.}}(Q_0)$ decreases near the Q_0^c region. As long as the system moves along $U_{\text{f.e.+t.h.}}(Q_0)$, nothing happens at Q_0^c. On the backward, a transition from $U_{\text{f.e.+t.h.}}(Q_0)$ to $U_{\text{g.s.}}(Q_0)$ is possible at Q_0^c; this is nothing but an electron transition from ψ_k^c to ψ_N^v, i.e., nonradiative capture of a free electron of the defect: a recombination has completed.

The following are emphasized on the CCD in Fig. 6.7.

(1) The direction of energy is always upward. The excitation occurs upward and the relaxation downward. In case I, for example, a point with energy ΔE_h larger than the minimum of $U_{\text{t.e.+f.h.}}(\bar{Q}_0)$ represents a state where an electron is bound to a defect with a binding energy E_e^{op}, a free hole moves with a kinetic energy ΔE_h, and the lattice distorts $\bar{Q}_0$.

(2) Any optical electronic transition occurs vertically starting from the minimum of the adiabatic potential of the initial state, except for hot luminescence. In case I, a transition from (t.e.+f.h.) to (f.e.+f.h.) corresponds to an electron emission with the optical ionization energy (optical depth) E_e^{op}. That from (g.s.) to (t.e.+f.h.) corresponds to a hole emission with

the optical ionization energy (optical depth) $E_{\mathrm{h}}^{\mathrm{op}}$ as if a hole is trapped by the defect in (g.s.). After an optical transition the lattice relaxation energy $E_{\mathrm{LR}} = \beta^2/2C$ is dissipated into the lattice vibration: $E_{\mathrm{e}}^{\mathrm{op}} + E_{\mathrm{h}}^{\mathrm{op}} = E_{\mathrm{g}} + 2E_{\mathrm{LR}}$.

(3) A nonradiative transition is possible if the lattice distorts to reach the crossing point Q_0^{c} or Q_0^{v} of the adiabatic potentials. A nonradiative capture of an electron (hole) starts from free electron (hole) states with a continuum $\psi_k^{\mathrm{c}}(\psi_k^{\mathrm{v}})$ to a localized state $\psi_1^{\mathrm{c}}(\psi_N^{\mathrm{v}})$. We can discuss successive captures of an electron and a hole and their correlation in one scheme (in Chapter 7).

(4) The lattice relaxation occurs after an electron capture and also after a hole capture. In case I, after an electron (hole) capture, the coordinate $Q_0(t)$ vibrates around $\bar{Q}_0$ ($Q_0 = 0$) with damping and an energy equal to the thermal depth $E_{\mathrm{e}}^{\mathrm{th}}$ ($E_{\mathrm{h}}^{\mathrm{th}}$) is dissipated over the crystal (multiphonon process: the average released phonon number is $E_{\mathrm{e}}^{\mathrm{th}}/\hbar\bar{\omega}_k$ ($E_{\mathrm{h}}^{\mathrm{th}}/\hbar\bar{\omega}_k$): $E_{\mathrm{e}}^{\mathrm{th}} + E_{\mathrm{h}}^{\mathrm{th}} = E_{\mathrm{g}}$.

(5) If we confine ourselves to capture and emission processes of a single carrier, we extract a part of the CC diagram (surrounded by a dotted line in Fig. 6.7). These parts are the same as Figs. 6.2b or 6.4b, respectively. For an electron capture and emission, a hole remains in a free state, so we can disregard a hole. In Fig. 6.7a, delete "f.h.", the single hatching which represents its continuum spectrum, and (g.s.). This gives us Fig. 6.2b. Thus, when the lattice relaxation energy E_{LR} is rather small compared to the band gap energy E_{g}, we can safely use a conventional CC diagram for a single carrier: Fig. 6.2 for a conduction electron and Fig. 6.4 for a valence hole.

So far, we have neglected the Coulomb interaction between an electron and a hole. In order to take account this exciton effects, we have to import the results of the effective mass approximation. Two other states are added:

(f.ex.): free exciton states with a single-continuum spectrum due to the kinetic energy of the center of mass motion; the lowest adiabatic potential is located $E_{\mathrm{B}} = m_{\mathrm{e}}^* m_{\mathrm{h}}^* e^4/2\varepsilon_0^2\hbar^2(m_{\mathrm{e}}^*+m_{\mathrm{h}}^*)$ below that of (f.e.+f.h.), where E_{B} is the exciton binding energy.

(t.ex.): a trapped exciton state; the adiabatic potential is located E_B below that of (t.e.+f.h.) or (f.e.+t.h.). The transition from (t.ex.) to (g.s.) is a capture by a charged defect: annihilation of a trapped exciton.

6.2.3 Far Deep into the Valence (Conduction) Band

The relative position of the adiabatic potentials for deep-level defects depends on the parameters $\varepsilon^c + \Delta_c$, $\varepsilon^v + \Delta_v$, E_g and E_{LR}. One of the most interesting cases would be a deep electron level buried into the valence band or a deep hole level buried into the conduction band, which is illustrated in Fig. 6.8. In these cases, the radiative transition from (t.e.+f.h.) or (f.e.+t.h.) to (g.s.) is not possible and only recombines via a thermally activated process over the potential barrier. A similar prohibition is applied to a bound exciton in which one of the carriers is deeply bound.

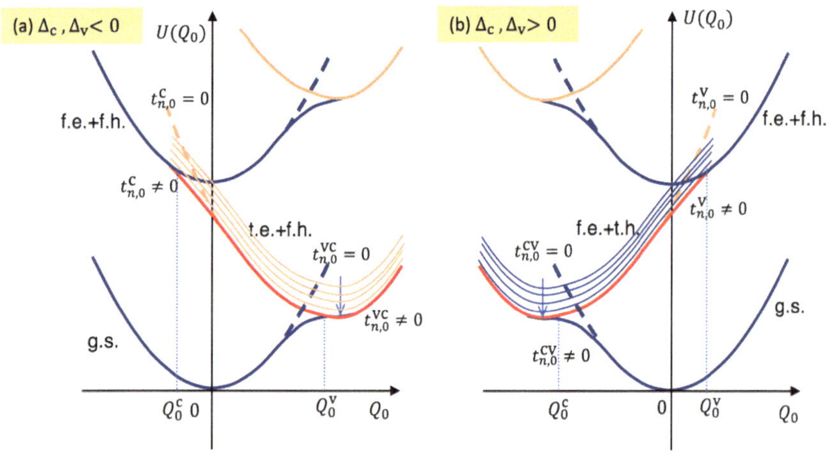

Figure 6.8 As the lattice deforms, (a) a deep electron level is sinking into the valence band, and (b) a deep hole level is sinking into the conduction band. Note that one electron energy $\varepsilon_l^\sigma(Q_0)$ in each case corresponds to Figs. 6.6(a), (b), respectively.

6.2.4 CC Diagram for Many Carriers

We have shown how to present the correct configuration coordinate diagrams for deep level defects even when the number of carrier changes by recombination. Using this correct CCD, we have shown that a nonradiative recombination of an electron and a hole at a deep-level defect consists of a set of subsequent two capture processes. During an electron (hole) capture, an electronic energy equal to the thermal depth E_e^{th} (E_h^{th}) is dissipated over the crystal ($E_e^{th} + E_h^{th} = E_g$).

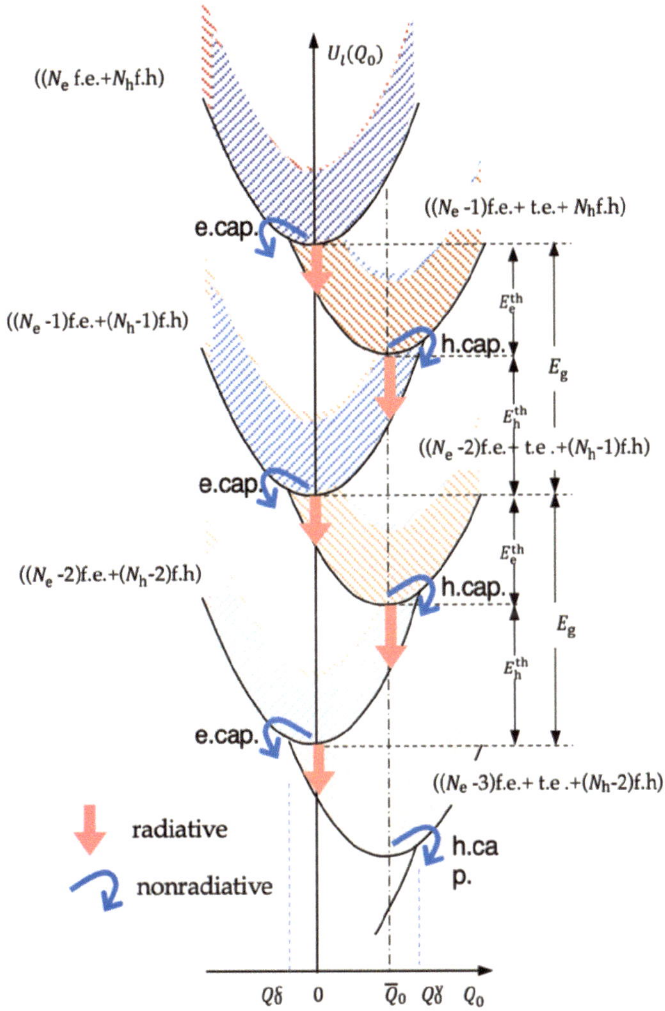

Figure 6.9 A typical configuration coordinate diagram for a deep-level defect with many carriers for case I: Δ_c, Δ_v <0 [3]. Hatchings represent continuum spectra of free carriers, instead of parabolas shown in Figs. 6.2 and 6.4.

In real situation, the number of carriers is not limited to one. When there are N_e electrons in the conduction band and N_h holes in the valence band, the adiabatic potentials with many carriers for the defect in Fig. 6.7 can be obtained by piling up those of (g.s.) and (t.e. + f.h.) or (f.e. + t.h.) with a sequence of the band gap energy E_g (Fig. 6.9) [6]. The interaction mode Q_0 is determined only by the charge distribution at the defect and is not affected by the number of free carriers. With using this CC diagrams, in the next chapter we will discuss the dynamics of the nonradiative multiphonon captures and recombination at deep-level defects.

6.2.5 Application to Typical Defects

Some defects in semiconductors show many charged states. The CCD at a charged defect can be obtained in a similar way. For example, at the EL2 center in GaAs, the replacement of three electronic states in Fig. 6.7a

(f.e. + f.h.) $\rightarrow$ EL2$^+$ + f.e. + f.h.

(t.e. + f.h.) $\rightarrow$ EL2^0 + f.h.

(g.s.) $\rightarrow$ EL2$^+$

gives a qualitatively correct CCD for EL2. The quantitative positions in energy between three states are obtained by experimental measurements from fitting with two quantities among four: the optical ionization energy (optical depth) E_e^{op}, E_h^{op} and the thermal depth E_e^{th}, E_h^{th} for an electron and/or a hole. The interaction mode Q_0 is defined between the minima of EL2$^+$ and EL2^0 configurations and the origin $Q_0 = 0$ refers to EL2$^+$. The absolute value of Q_0 cannot be evaluated by energy spectroscopies except for ENDOR; hence, it usually remains a conceptual quantity.

If a defect shows three types of charged states, for example, EL2^0, EL2$^+$ and EL2^{++} in GaAs, at least six electronic states must be considered: for example,

(2f.e. + 2f.h.) $\rightarrow$ EL2^{++} + 2f.e. + 2f.h.

(t.e. + f.e. + 2f.h.) $\rightarrow$ EL2$^+$ + f.e. + 2f.h.

(f.e. + f.h.) $\rightarrow$ EL2^{++} + f.e. + f.h.

(2t.e. + 2f.h.) $\rightarrow$ EL2^0 + 2f.h.

(t.e. + f.h.) $\rightarrow$ $EL2^{+}$ + f.h.

(g.s.) $\rightarrow$ $EL2^{++}$.

The adiabatic potentials should be presented in two configurational coordinate space $Q_{+,0}$ and $Q_{++,+}$. Here the interaction mode $Q_{+,0}$ $Q_{++,+}$) is defined in the direction between two minima of $EL2^{+}$ and $EL2^{0}$ ($EL2^{++}$ and $EL2^{+}$).

6.2.6 Several Localized Electrons

So far, we have discussed CCD for such defects which attract a single electron or hole. In case of several localized electrons at a defect, there are three methods to calculate the adiabatic potentials.

(1) If the defect couples only the conduction (valence) band, introduce an effective Hamiltonian which consists of n_e extra electrons (holes) and the lattice vibrations. For the electron case,

$$H_{\text{eff}} = H_{\text{e}}(\{\boldsymbol{r}_i\}) + V_{\text{eL}}\big(\{\boldsymbol{r}_i\},\{Q_{k,\sigma}\}\big) + H_{\text{L}}\big(\{Q_{k,\sigma}\}\big). \quad (6.28)$$

$$H_{\text{e}}(\{\boldsymbol{r}_i\}) = \sum_{i=1}^{n_e}\left(\frac{p_i^2}{2\tilde{m}_{\text{e}}} + V_{\text{def}}(\{\boldsymbol{r}_i\})\right) + \sum_{i}^{n_e}\sum_{j(\neq i)}^{n_e} V_{\text{ee}}(\{\boldsymbol{r}_i - \boldsymbol{r}_j\}), \quad (6.29\text{a})$$

$$V_{\text{eL}}\big(\{\boldsymbol{r}_i\},\{Q_{k,\sigma}\}\big) = -\sum_{k,\sigma}\sum_{i=1}^{n} A^{\text{c}}_{k,\sigma}\, Q_{k,\sigma} \exp(i\boldsymbol{k}\cdot\boldsymbol{r}_i). \quad (6.29\text{b})$$

Here $V_{\text{der}}(\{\boldsymbol{r}_i\})$ denotes the defect potential and $V_{\text{ee}}(\{\boldsymbol{r}_i - \boldsymbol{r}_j\})$ the electron–electron interaction. If the effective mass approximation can be applied, that is, states can be expressed by a linear combination of the Wannier states of the conduction band, $\tilde{m}_e$ can be replaced by the electron effective mass m_e^* and the form of the electron–lattice interaction Eqs. (2.37–2.39) can be used.

(2) The second way is to formally expand the adiabatic potentials of the l-th states (includes both carriers consistently) with respect to the lattice distortion $\{\Delta R_j\}$ or the normal mode $\{Q_{k,\sigma}\}$ which are defined in the ground state.

$$U_l(\{\Delta R_j\}) = U_l(\{\Delta R_j = 0\}) + \sum_j \left(\frac{\partial U_l}{\partial \Delta R_j}\right)_{\{\bar{R}\}_g} \Delta R_j + \cdots$$

$$= U_l(\{0\}) + \sum_{k,\sigma} a^l_{k,\sigma}\, \tilde{Q}_{k,\sigma} + \cdots \tag{6.30}$$

The second term expresses the electron–lattice interaction. The coefficients $a^l_{k,\sigma}$ and $U_l(\{0\})$ are regarded as parameters, with which one can calculate the l-th adiabatic potential and fit it to the experimental data, paying attention to the point symmetry of the system. This is usually done in the Jahn–Teller system.

(3) The third way is to perform an ab initio calculation, which directly gives us the adiabatic potentials. Thanks to the rapid progress in computer ability, the first-principles calculation of electronic and phononic structures in condensed materials can be performed for not only crystal systems but also defects systems. In calculations for the defect systems where the Bloch theorem cannot be applied any more, a choice of super cell of ample size is necessary. Figure 6.10 shows a recent result on first-principles calculation on Li_{Zn} substitutional impurity in ZnO [7].

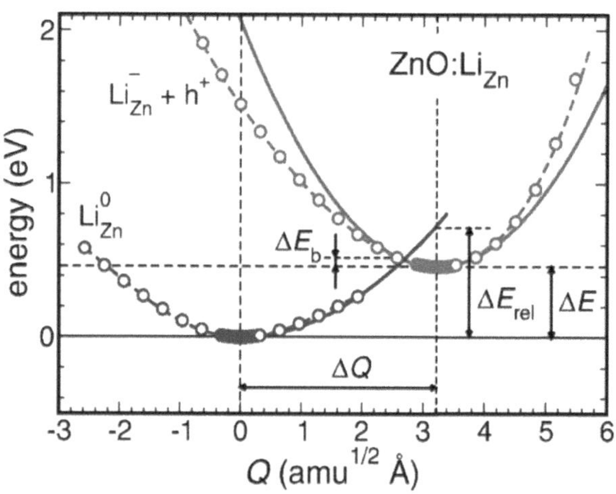

Figure 6.10 Calculated 1-dimensional configuration coordinate diagram for a hole capture at the LiZn defect in ZnO using first-principles calculation. Reproduced with permission from Ref. [7].

References

1. Stoneham, A. M., *Theory of Defects in Solids* (Oxford, 1975).
2. Watkins, G. D., *Proc. Defects in Semiconductors* 1972, 228; *Physica* **117B&11**, (1983) 9.
3. Shinozuka, Y., *Jpn. J. Appl. Phys.*, **32** (1993) 4560.
4. Henry, C. H., and Lang, D., *Phys. Rev. B*, **15** (1977) 989.
5. Toyozawa, Y., *Solid-State Electron.*, **21** (1978) 1313.
6. Shinozuka, Y., *J. Phys. Soc. Jpn.*, **51** (1982) 2852.
7. Alkauskas, A., Yan, Q., and Van de Walle, C. G., *Phys. Rev. B*, **90** (2014) 075202.

Chapter 7

Nonradiative Multiphonon Capture and Recombination Processes

We have seen in Chapter 6 that when the electron–lattice interaction is strong at a deep level defect in semiconductors, there arise level crossings between adiabatic potentials for a localized state and delocalized states. Then nonradiative carrier captures and emissions become possible. In this chapter, we study the dynamics of the nonradiative recombination at deep-level defects using appropriate configuration coordinate diagrams for many carriers introduced in the previous chapter. The energy conversions from carriers to the lattice vibrations and the feedback mechanism are discussed.

7.1 Nonradiative Multiphonon Recombination

Some deep-level defects in semiconductors have been considered to act as nonradiative recombination centers and limit the efficiency of light-emitting devices. They are called killer centers [1]. There are two major nonradiative processes: One is the Auger process and the other is the multiphonon process. In the Auger

Electron–Lattice Interactions in Semiconductors
Yuzo Shinozuka

ISBN 978-981-4800-96-9 (Hardcover), 978-1-003-02346-3 (eBook)
www.jennystanford.com

process, an electron–hole (e–h) pair is annihilated in exchange for an excitation of another free carrier [2]. The e–h pair energy ($\sim E_g$) is transformed to the kinetic energy of a free carrier, which is later converted to phonons. This Auger process is probable in high carrier density case and governs the lifetime of carriers. In the multiphonon recombination process at defects, carriers recombine with emitting phonons: The electronic energy is converted to the lattice vibration energy. This process is very serious in light-emitting devices since it limits the efficiency and is also considered to be a trigger of defect reactions (phonon-kick mechanism), resulting in structural degradation of devices [3–7]. In the following, we discuss the whole process starting from a carrier capture by a deep-level defect and succeeding transient induced lattice vibrations using appropriate configuration coordinate diagrams (CCD) for many carriers. Although we use the name "multiphonon," most of the following analysis is based on a classical treatment of the lattice motion. The following in this chapter are based on [8].

7.1.1 Configuration Coordinate Diagram

Let us consider a point defect that has two charge states, D^0 and D^-, as an example. Generalization is possible and will be discussed later. Figure 7.1 shows a typical example of the configuration coordinate diagram (CCD) [8]. In the figure, [N_e, N_h, D^0] ([N_e, N_h, D^-]) denotes a state with N_e free electrons in the conduction band, N_h free holes in the valence band and D^0 (D^-) the defect charge. The adiabatic potentials $U(Q_0)$ for [N_e, N_h, D^0] all have a minimum at $Q_0 = 0$ and are located with an equal vertical distance the band gap energy E_g, and $U(Q_0)$ for [N_e, N_h, D^-] all have a minimum at $\bar{Q}_0$ and are separated from each other by E_g. Note that each adiabatic potential is accompanied by a set of $N_e \times N_h$ multiple continuum states representing the kinetic energies of free carriers, however, we do not indicate them in Fig. 7.1 to avoid graphic complexity. Based on the results and discussion in Chapters 3 and 6, we can derive the following:

(a) The ordinate Q_0 can be regarded as the interaction mode Q_{int} that connects the stable configurations for two electronic states, for example D^+ and D^0. The origin $Q_0 = 0$ is taken for D^0

displacements around the defect and is a linear combination of many normal modes in general.

(b) In the CCD, each single point represents the charge of the defect and the kinetic energies of free carriers, and the lattice elastic energy stored around the defect as a function of the lattice distortion Q_0. The direction of increasing energy is always upward for electrons, holes, and the elastic lattice.

(c) The relative position between $U(Q_0)$'s for [N_e, N_h, D^0] and [N_e, N_h, D^-] depends on the parameters of the defect: ε_0, Q_0^c, Q_0^v, and E_{LR} (Table 7.1). Note that the subscripts and the superscripts in $E_{th}^{e(h)}$ and $E_{op}^{e(h)}$ are interchanged from Chapter 6.

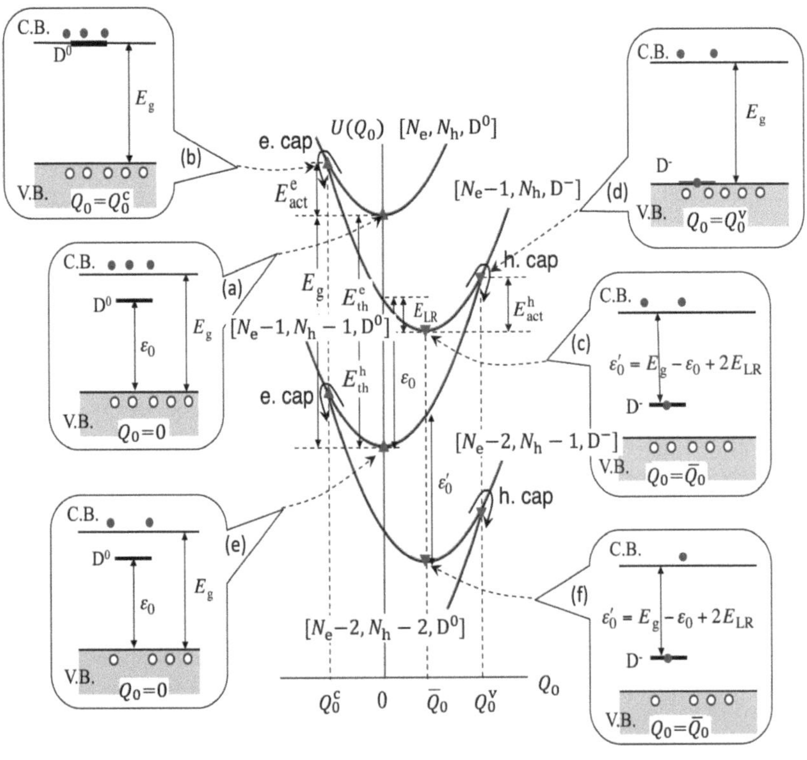

Figure 7.1 A typical configuration coordinate diagram (CCD) for a deep-level defect with many carriers. (a)–(f) One-electron energy and carrier occupation at the configurations $Q_0 = 0$, Q_0^c, Q_0^v and $\bar{Q}_0$, respectively. Reproduced with permission from Ref. [8].

Table 7.1 Relationship among characteristic energies of a defect: E_g, ε_0, E_{LR}

Reference carrier	Electron	Hole
Optical depth	$E_{op}^{e} = E_g - \varepsilon_0 + 2E_{LR} = \varepsilon_0'$	$E_{op}^{h} = \varepsilon_0$
Thermal depth	$E_{th}^{e} = E_g - \varepsilon_0 + E_{LR}$	$E_{th}^{h} = \varepsilon_0 - E_{LR}$
Activation energy for nonradiative capture	$E_{act}^{e} = (E_g - \varepsilon_0)^2/4E_{LR}$	$E_{act}^{h} = (\varepsilon_0 - 2E_{LR})^2/4E_{LR}$

(d) At the crossing point Q_0^c (Q_0^v) between two adiabatic potentials where the (transient) optical depth of a trapped electron (hole) is zero, a nonradiative transition of an electron (hole) is possible. This electron (hole) transition is a capture from free electron (hole) continuum states to a single localized state.

(e) Once a nonradiative electron capture occurs at Q_0^c, which follows the lattice relaxation, an electronic energy equal to E_{th}^{e} of a trapped electron is converted to the lattice vibration energy. The average number of generated phonons is $S = E_{th}^{e}/\hbar\omega$ (multiphonon process).

(f) At the stable configuration $\bar{Q}_0$ for D^0, where the defect is occupied by an electron, the lattice elastic energy stored around the defect is E_{LR}.

(g) Once a nonradiative hole capture occurs at Q_0^v, which follows the lattice relaxation, an energy equal to the thermal depth $E_{th}^{h} = E_g - E_{th}^{e}$ of a trapped hole is converted to the lattice vibration energy. Thus, during a pair of successive electron and hole captures, the electronic energy $E_g = E_{th}^{e} + E_{th}^{h}$ is converted to phonons: A nonradiative multiphonon recombination is completed.

(h) Optical electronic transitions may also take place vertically in the CCD according to the Franck–Condon principle, then the optical depth E_{op}^{e} of a trapped electron at the stable configuration $\bar{Q}_0$ is given by $E_{th}^{e} + E_{LR}$ and the optical depth E_{op}^{e} of a trapped hole at the stable configuration $Q_0 = 0$ is given by $E_{th}^{h} + E_{LR}$.

Our next task is to find the probability of each nonradiative carrier capture.

7.1.2 First Carrier Capture

Consider the p-type condition, for example, where the defect is usually D^0. Let us start with the first electron capture process of the defect (D^0 + free electron $\rightarrow$ D^-) under thermal lattice vibration around $Q_0 = 0$. It has been shown theoretically that the nonradiative multiphonon electron (minority) capture rate of a defect per unit time is given by

$$W_{\mathrm{c}}^{\mathrm{e}} = \frac{\bar{\omega}}{2\pi} \exp\left(-\frac{E_{\mathrm{act}}^{\mathrm{e}}}{k_{\mathrm{B}}T}\right) P_{\mathrm{c}}^{\mathrm{e}}. \tag{7.1}$$

Here, the first factor represents the attempt frequency indicating the number of times the lattice vibrates per time, where $\bar{\omega}$ is the mean angular frequency. The second factor presents the activation $Q_0 = 0 \rightarrow Q_0^{\mathrm{c}}$ of the lattice and the third factor $p_{\mathrm{c}}^{\mathrm{e}}$ is the probability that the defect captures an incoming electron at $Q_0 = Q_0^{\mathrm{c}}$. Note that the electronic transition is possible from one of free-states to a localized state when $Q_0(t)$ backwardly crosses Q_0^{c}.

Sumi [9–11], and Kayanuma and Fukuchi [12] studied theoretically the capture probability of one free electron in the conduction band. Using the detailed balance of electronic transitions between the localized state and free states, they showed that $P_{\mathrm{c}}^{\mathrm{e}}$ does not depend so much on the nature of the defect and is given by

$$P_{\mathrm{c}}^{\mathrm{e}} = 1/n_T^{\mathrm{e}} V, \tag{7.2}$$

where

$$n_{\mathrm{T}}^{\mathrm{e}} = \frac{1}{V} \sum_k \exp\left(-\frac{\varepsilon_k}{k_{\mathrm{B}}T}\right), \tag{7.3}$$

is the effective carrier density at temperature T and V is the volume of the semiconductor. For a parabolic energy band ε_k of a carrier with the effective mass m_i^*, n_{T}^i is shown to be $2(m_i^* k_{\mathrm{B}} T / 2\pi\hbar^2)^{3/2}$ (i = e or h). If there are N_{e} free electrons in the conduction band, the probability is then given by

$$P_c^e = 1 - (1 - 1/n_T^e V)^{N_e}. \tag{7.4}$$

When the carrier density $n^e \equiv N_e/V$ is not very high, P_c^e is approximately given by $P_c^e \approx N_e/n_T^e V = n^e/n_T^e$ and it tends to 1 for the high n^e limit. When E_{act}^e (E_{act}^h) is less than the thermal energy $k_B T$, a nonradiative capture of an electron (hole) would be probable by the defect in p-type (n-type), which starts from the activation $Q_0 = 0 \to Q_0^c$ ($Q_0 = \bar{Q}_0 \to Q_0^v$).

7.1.3 Induced Transient Lattice Vibration

After the defect has captured a free electron, each normal lattice mode $q_k(t)$ starts to vibrate around a new equilibrium $q_k = \bar{q}_k$ with its own frequency: $q_k(t) \propto \cos\omega_k t$ (see Chapter 3). As for the interaction mode Q_0, which consists of many normal modes with different frequencies, dephasing occurs and $Q_0(t)$ shows a damping vibration with mean frequency $\bar{\omega}$. If we assume that the distribution of $\{\omega_k^2 \bar{q}_k\}$ is given by a Gaussian with a width $\Delta\omega$, the decay time is $\tau \sim 2\pi/\Delta\omega$. Figures 7.2a-1 and 7.2c-1 show the typical time dependence of $Q_0(t)$. For the first capture process, the thermal energy E_{act}^e is necessary to activate the lattice at the defect ($Q_0 = 0 \to Q_0^c$), and during the following relaxation process ($Q_0 = Q_0^c \to \bar{Q}_0$), $E_{act}^e + E_{act}^h$ is transferred to the phonon system. As a result, the difference, E_{th}^e, which is equal to the thermal depth of a localized electron, is converted to the lattice vibration energy.

7.1.4 Second Capture

After the transient vibration has damped out ($t \gg \tau$), a nonradiative hole capture by the defect (D^- + free hole $\to D_0$) needs the activation energy E_{act}^h ($Q_0 = \bar{Q}_0 \to Q_0^v$). The probability per time is given by Eq. (7.1) with suffix e being replaced by h.

It has been suggested that the transient vibration induced by carrier capture enhances the probability of the subsequent capture of the opposite carrier [8]. If lattice $Q_0(t)$ crosses Q_0^v during the induced vibration triggered by an electron capture, the probability of an athermal hole capture per crossing at Q_0^v is given by

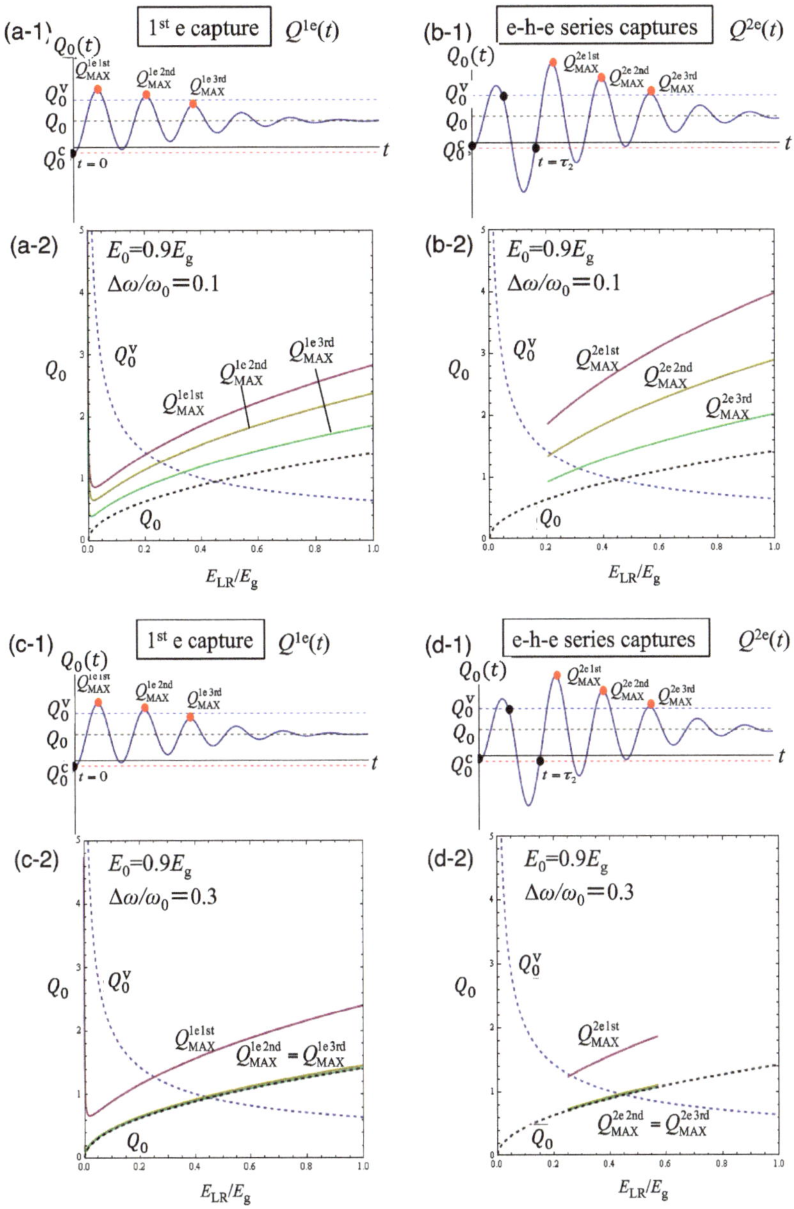

Figure 7.2 The maximum lattice displacement Q_{MAX} during the transient vibration induced by single e capture and a series e–h–e captures, as a function of E_{LR}/E_g for fixed $E_0 = 0.9E_g$ case. Reproduced with permission from Ref. [8]. Here $\omega_0 = \overline{\omega}$.

$$P_{\mathrm{c}}^{\mathrm{h}} = 1 - \left(1 - 1/n_{\mathrm{T}}^{\mathrm{h}}V\right)^{N_{\mathrm{h}}} \sim N_{\mathrm{h}}/n_{\mathrm{T}}^{\mathrm{h}}V = n_{\mathrm{h}}/n_{\mathrm{T}}^{\mathrm{h}}. \tag{7.5}$$

In the case shown in Fig. 7.2(a-1), $Q_1(t)$ crosses Q_0^{v} twice, which means two chances for the athermal capture. Figure 7.3 shows the number of times $Q_0(t)$ crosses during the lattice relaxation caused by an electron capture starting from Q_0^{c}. Thus, if $E_{\mathrm{act}}^{\mathrm{e}}$ + $E_{\mathrm{th}}^{\mathrm{e}} > E_{\mathrm{act}}^{\mathrm{h}}$, $\tau << 2\pi/\overline{\omega}$ (hence, $\Delta\omega/\overline{\omega} << 1$) and the hole density n_{h} is very high, athermal hole captures would possibly occur successively without activation energy $E_{\mathrm{act}}^{\mathrm{h}}$. Once it occurs, the stable position of the lattice is again $Q_0 = 0$, and then $Q_0(t)$ subsequently shows new damping vibration around $Q_0 = 0$ (see Fig. 3.3(b) in Chapter 3). During the relaxation, the electronic energy equal to the thermal depth $E_{\mathrm{th}}^{\mathrm{h}} = E_{\mathrm{g}} - E_{\mathrm{th}}^{\mathrm{e}}$ of a hole (as if a hole is localized in D^0) is dissipated to lattice vibration energy. As a result of successive captures of an electron and a hole, the electronic energy $E_{\mathrm{g}} = E_{\mathrm{th}}^{\mathrm{e}} + E_{\mathrm{th}}^{\mathrm{h}}$ is converted to lattice vibration energy; a nonradiative recombination has completed.

If the lattice $Q_0(t)$ crosses Q_0^{c} during the induced vibration triggered by a hole capture, it would enhance the probability of the second electron capture; hence, it would affect all the following carrier capture processes.

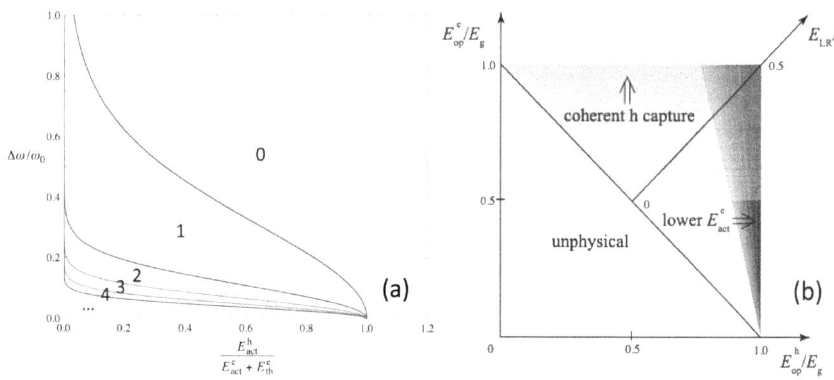

Figure 7.3 (a) How many times $Q_0(t)$ crosses Q_{h} during the transient vibration induced by an electron capture starting from Q_0^{c}. Reproduced with permission from Ref. [8]. (b) Phase diagram indicates the direction of lower activation energy $E_{\mathrm{act}}^{\mathrm{e}}$ for a minority electron capture and higher probability for coherent hole capture in p-type semiconductors as functions of the relative optical depth $E_{\mathrm{op}}^{\mathrm{e}}$ and $E_{\mathrm{op}}^{\mathrm{h}}$ of a deep-level.

7.2 Simulation of Successive Captures

7.2.1 Model

Let us simulate the time evolution of $Q_0(t)$ for $P_c^e{\sim}1$ and $P_c^h{\sim}1$ case. If several free electrons and free holes were nonradiatively captured by a defect, each occurs at time $\tau_{e1} < \tau_{h1} < \tau_{e2} < \tau_{h2} < \cdots < \tau_{eN} < \tau_{hN}$, the time evolution of the induced lattice vibration is given by [8]

$$Q_0(t) = Q_0^c \exp\left[-\frac{\Delta\omega^2(t-\tau_{e1})^2}{4}\right]\cos\bar{\omega}(t-\tau_{e1})$$
$$-\bar{Q}_0\sum_{j=2}^{N}\left\{\exp\left[-\frac{\Delta\omega^2\left(t-\tau_{ej}\right)^2}{4}\right]\cos\bar{\omega}\left(t-\tau_{ej}\right)-1\right\}\theta\left(t-\tau_{ej}\right)$$
$$+\bar{Q}_0\sum_{j=1}^{N}\left\{\exp\left[-\frac{\Delta\omega^2\left(t-\tau_{hj}\right)^2}{4}\right]\cos\bar{\omega}\left(t-\tau_{hj}\right)-1\right\}\theta\left(t-\tau_{hj}\right). \tag{7.6}$$

Here, $\theta(t)$ is a step function [$\theta(t) = 0$ for $t < 0$ and $\theta(t) = 1$ for $t > 0$]. The capture time τ_{ej} (τ_{hj}) cannot take an arbitrary value, but should be one of the times when $Q_0(t)$ crosses Q_0^c (Q_0^v) with the probability P_c^e (P_c^h): $Q_0(\tau_{ej}) = Q_0^c$ and $Q_0(\tau_{hj}) = Q_0^v$. Thus, the whole process turns out to be a cooperative probability process.

Once $Q_0(t)$ will no longer cross Q_0^c or Q_0^v, the series of coherent captures has been terminated. Figure 7.2 shows typical examples of the maximum lattice displacement Q_{MAX} during the transient vibration induced by carrier capture(s). Since we are interested in the largest value of Q_0, we indicate here only a rapid process in which each subsequent capture occurs at the first crossing at Q_0^c (Q_0^v). In Figs. 7.2a-2 and 7.2c-2, $Q_{MAX}^{1e\,n\text{th}}$ denotes the maximum value during the n-th vibrational period after the first electron capture. When $\Delta\omega/\bar{\omega} \neq 0$, $Q_{MAX}^{1e\,1\text{st}} > Q_{MAX}^{1e\,2\text{nd}} > Q_{MAX}^{1e\,3\text{rd}} > \ldots$ because of damping due to dephasing. If $Q_{MAX}^{1e\,n\text{th}}$ is larger than Q_h, shown by the dotted line in Fig. 7.2b, an athermal hole capture is possible during the n-th vibration period. $Q_{MAX}^{2e\,n\text{th}}$ denotes the maximum lattice displacement during the n-th vibration period after the second electron capture triggered by the first electron capture and first hole capture that both occurred at the first crossing. As can be seen in the case of $\Delta\omega/\bar{\omega} = 0.1$, $Q_{\mathrm{MAX}}^{2\mathrm{e}\,n\mathrm{th}} > Q_{\mathrm{MAX}}^{1\mathrm{e}\,n\mathrm{th}}$, where inflation of induced lattice vibrations would be possible through a series of athermal captures. On the other hand, when $\Delta\omega/\bar{\omega} = 0.3$, $Q_{\mathrm{MAX}}^{2\mathrm{e}\,n\mathrm{th}} \sim Q_{\mathrm{MAX}}^{1\mathrm{e}\,n\mathrm{th}}$, then such inflation does not occur even when a series of athermal captures occurs.

7.2.2 Inflation of Vibration

We have seen that if $E_{\mathrm{act}}^{\mathrm{e}} + E_{\mathrm{th}}^{\mathrm{e}} > E_{\mathrm{act}}^{\mathrm{h}}$ and both carrier densities are high enough, i.e., $P_{\mathrm{c}}^{\mathrm{e}} \sim 1$ and $P_{\mathrm{c}}^{\mathrm{h}} \sim 1$, a series of coherent nonradiative carrier aptures are possible. The probability per time would be given by

$$W_{\mathrm{c}}^{\mathrm{e-h-e-\cdots-h}} = W_{\mathrm{c}}^{\mathrm{e}} = \frac{\omega_0}{2\pi}\exp\left(-\frac{E_{\mathrm{act}}^{\mathrm{e}}}{k_B T}\right) P_{\mathrm{c}}^{\mathrm{e}} \times P_{\mathrm{c}}^{\mathrm{h}} \times P_{\mathrm{c}}^{\mathrm{e}} \times \ldots \times P_{\mathrm{c}}^{\mathrm{h}}. \tag{7.7}$$

Here, the Arrhenius dependence originates only from the first minority capture. In typical III–V semiconductors where $m_i^* \sim 0.2m_{\mathrm{e}}$, $n_{\mathrm{T}}^i = 2(m_i^* k_{\mathrm{B}}T/2\pi\hbar^2)^{3/2} \sim 2.0 \times 10^{18}\ \mathrm{cm}^{-3}$ for $T = 300$ K. In active light-emitting devices, $n^i \sim n_{\mathrm{T}}^i$ and hence both $P_{\mathrm{c}}^{\mathrm{e}} \sim 1$ and $P_{\mathrm{c}}^{\mathrm{h}} \sim 1$ can be realized. If $\Delta\omega/\bar{\omega} < 0.3$, an inflation of the lattice vibration caused by a series of coherent captures would overcome the potential barrier E_{t} for the defect reaction; this would be a plausible mechanism of "phonon kick," which will be discussed in Chapter 9. If $\Delta\omega/\bar{\omega} > 0.3$, on the other hand, the inflation does not occur, because the energy dissipation is so rapid in a vibration period that the "local" vibration energy does not accumulate markedly (Fig. 7.3).

The following should be noted:

(1) In n-type semiconductors, the first hole capture starts from D^- with the equilibrium lattice around $\bar{Q}_0$. It can be similarly discussed with a suitable interchange between $Q_0 = 0$ and $\bar{Q}_0$.

(2) In device physics, the capture cross section σ^i of a carrier (i = e, h) is widely used. Its relation to the present capture probability W_{c}^i of a defect is $W_{\mathrm{c}}^i = n_{\mathrm{T}}^i v_{\mathrm{T}}^i \sigma^i$, where $v_{\mathrm{T}}^i = (3k_{\mathrm{B}}T/m_i^*)^{1/2}$ is the thermal velocity.

(3) In the present analysis, we have restricted ourselves to the case where both the optical depth ε_0 and ε_0' are less than E_g (Table 7.1). If ε_0 (ε_0') is larger than E_g, the localized state is resonant with the conduction (valence) band. Then, the electron (hole) capture process is nothing but extrinsic self-trapping.

7.2.3 Summary

We have discussed possibility of a series of coherent nonradiative carrier captures by a defect in semiconductors. It can be summarized as follows:

(a) $P_c^e << 1$ and $P_c^h << 1$

A single nonradiative e (h) capture independently occurs through a thermal activation from $Q_0 = 0 \to Q_0^c$ ($Q_0 = \bar{Q}_0 \to Q_0^v$) with the probability per time

$$W_c^e = \frac{\bar{\omega}}{2\pi} \exp\left(-\frac{E_{act}^e}{k_B T}\right) P_c^e, \tag{7.8a}$$

$$W_c^h = \frac{\bar{\omega}}{2\pi} \exp\left(-\frac{E_{act}^h}{k_B T}\right) P_c^h. \tag{7.8b}$$

(b) $P_c^e << 1$ and $P_c^h \sim 1$: p-type

A set of successive nonradiative e capture and following h capture is possible, which started by a thermal activation from $Q_0 = 0 \to Q_0^c$ with the probability per time.

$$\frac{\bar{\omega}}{2\pi} \exp\left(-\frac{E_{act}^e}{k_B T}\right) P_c^e \; P_c^h, \tag{7.9}$$

(c) $P_c^e \sim 1$ and $P_c^h << 1$: n-type

A set of successive nonradiative h capture and following e capture is possible, which started by a thermal activation from $Q_0 = \bar{Q}_0 \to Q_0^v$ with the probability per time,

$$\frac{\bar{\omega}}{2\pi} \exp\left(-\frac{E_{act}^h}{k_B T}\right) P_c^h \; P_c^e, \tag{7.10}$$

(d) $P_c^e \sim 1$ and $P_c^h \sim 1$: high injection of both carriers

n-pairs of coherent nonradiative e and h captures are possible, which started either by a thermal activation from $Q_0 = 0 \rightarrow Q_0^c$ with the probability per time

$$\frac{\bar{\omega}}{2\pi} \exp\left(-\frac{E_{act}^e}{k_B T}\right) \left(P_c^h \; P_c^e\right)^n \tag{7.11a}$$

or from $Q_0 = \bar{Q}_0 \rightarrow Q_0^v$ with the probability per time,

$$\frac{\bar{\omega}}{2\pi} \exp\left(-\frac{E_{act}^h}{k_B T}\right) \left(P_c^e \; P_c^h\right)^n \tag{7.11b}$$

We have shown that the possibilities of coherent nonradiative carrier capture and inflation of the induced lattice vibration depend critically on the relative width of frequency distribution $\Delta\omega/\bar{\omega}$. Then, the details of the interaction mode Q_0, i.e., whether the mode contains the local mode(s) at the defect, are important. Studies on the interaction modes at typical defects in III–V and II–VI semiconductors are necessary for the progress of defect controls in these semiconductors.

References

1. Gold, R. D., and Weisberg, L. R., *Solid-State Electron*, **7** (1964) 811.
2. Landsberg, P. T., and Adams, M.J., *J. Luminescence*, **7** (1973) 3.
3. Dean, P. J., and Choyke, W. J., *Adv Phys.*, **26** (1977) 1.
4. Lang, D. V., *Annu. Rev. Mater. Sci.*, **12** (1982) 377.
5. Englman, R., *Non-Radiative Decay of Ions and Molecules in Solids* (North Holland, 1979).
6. Hayashi, I., *Proc. 15th Int. Conf. Physics Semiconductors*, Kyoto, 1980, **21** (1966) 1663.
7. Wada, K., and Pang, S. W. (eds.), *Defects in Optoelectronic Materials*, Gordon and Breach Science Publishers, 2001.
8. Shinozuka, Y., Wakita, W., and Suzuki, K., *Jpn. J. Appl. Phys.*, **51** (2012) 11PC03.

9. Sumi, H., *J. Phys. Soc. Jpn.*, **49** (1980) 1701.
10. Sumi, H., *Phys. Rev. Lett.*, **47** (1981) 1333.
11. Sumi, H., *Phys. Rev. B*, **27** (1983) 2374.
12. Kayanuma, Y., and Fukuchi, S., *J. Phys. Soc. Jpn.*, **53** (1984) 1869.

Chapter 8

Symmetry Breaking Processes

So far, we have learned that the stable configuration of ions (atoms) in semiconductors may change by electronic excitations: photonic excitations and carrier doping. Strong electron–lattice interaction manifests itself in association with localization of carrier. In this chapter, we study the rearrangements of atoms in real space, in particular, symmetry breaking processes.

8.1 Jahn–Teller Effect

In Section 4.1, using a continuum model we have studied an electron localization assuming a spherical symmetry. If an electron becomes strongly localized so as the extension a of the wave function is the order of the lattice constant d, the electron feels the local arrangement of ions (atoms) and also feels other valence electrons around it. There we must explicitly consider the geometric defect structure, its symmetry, and the interaction among localized electrons to discuss the lattice distortions generated by electronic transitions.

Several point defects in semiconductors have been found to show the off-center instability, where the central atom at the defect is displaced to one of the off-center positions [1–3]. Semi-insulating GaAs shows a bistability of the configuration

Electron–Lattice Interactions in Semiconductors
Yuzo Shinozuka

ISBN 978-981-4800-96-9 (Hardcover), 978-1-003-02346-3 (eBook)
www.jennystanford.com

relating to a midgap state called EL2. The mechanism is explained theoretically by Chadi and Chan using a first-principles calculation [4]. They showed the bistability comes from the on-off center instability of an antisite As atom. Figure 8.1 shows the structure of EL2 and the off-center distortion, experimentally observed by STM measurements by Hida et al. [5]. A similar metastability has been found in DX-center in AlGaAs [6] and in a substitutional Si impurity [7, 8].

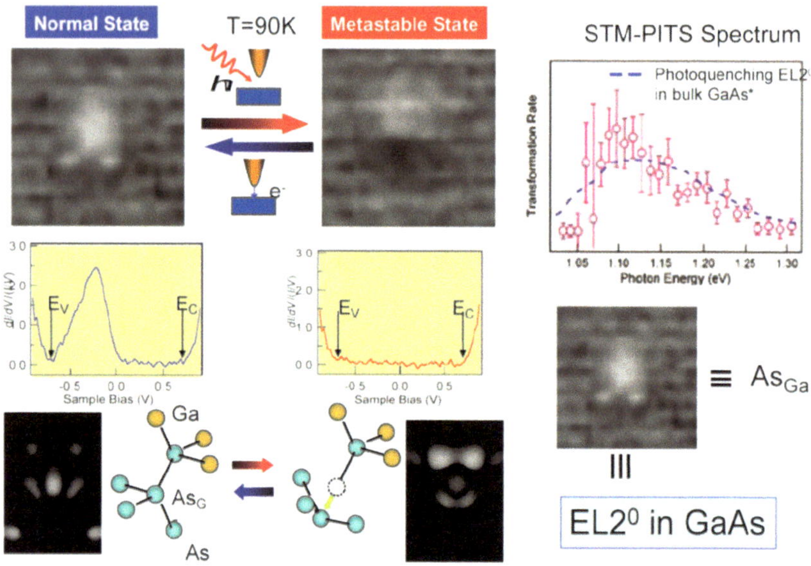

Figure 8.1 STM images of EL2 center in GaAs. As the charge of EL2 changes by photo excitation or carrier injection, the stable position of antisite As atom moves between on center and off-center. Courtesy of K. Maeda.

These off-center instabilities can be understood in terms of the Jahn–Teller (J–T) effect, which has been first proposed to discuss the stability of molecular systems [9]. If the atomic structure of a point defect possesses a certain point-symmetry, the localized electronic states are often degenerate. The J–T theory tells us that for degenerate electronic states the structure is unstable and molecules are always deformed to lower the symmetry.

Let us consider a point defect in a semiconductor, whose point symmetry is characterized by a symmetry group denoted by **G** (Fig. 8.2) [10]. The adiabatic Hamiltonian of the system is given by

$$H = H_e(\{r\}) + V_{eL}(\{r\}, \{\Delta R\}) + U_L(\{\Delta R\}). \tag{8.1}$$

Here $\{r\}$ denotes a set of the position of electrons $\{r_1, r_2, ...\}$ and $\{\Delta R\}$ represents a set of displacements $\{\Delta R_1, \Delta R_2, ...\}$ of ions around the defect. We must distinguish the relevant electrons and the other valence electrons that are combined with all ions to form the lattice medium. Hence, we refer "atom", rather than ion hereafter. In Eq. (8.1), $U_L(\{\Delta R\})$ is the elastic energy of the displacements of atoms. The set of displacements $\{\Delta R\}$ can be transformed into a set of normal vibration modes $\{Q_\Gamma\}$ in the irreducible representation Γ of $\mathbf{G}$. The electron–lattice interaction term $V_{eL}(\{r\}, \{\Delta R\}) = V_{eL}(\{r\}, \{Q_\Gamma\})$ can be expanded as a power series of ΔR_j's or Q_Γ's.

$$V_{eL}(\{r\}, \{\Delta R\}) = V_{eL}(\{r\}, \{\Delta R = 0\}) + \sum_j \frac{\partial V_{eL}}{\partial \Delta R_j} \Delta R_j + \cdots$$

$$V_{eL}(\{r\}, \{Q_\Gamma\}) = V_{eL}(\{r\}, \{Q_\Gamma = 0\}) + \sum_j \frac{\partial V_{eL}}{\partial Q_\Gamma} Q_\Gamma + \cdots. \tag{8.2}$$

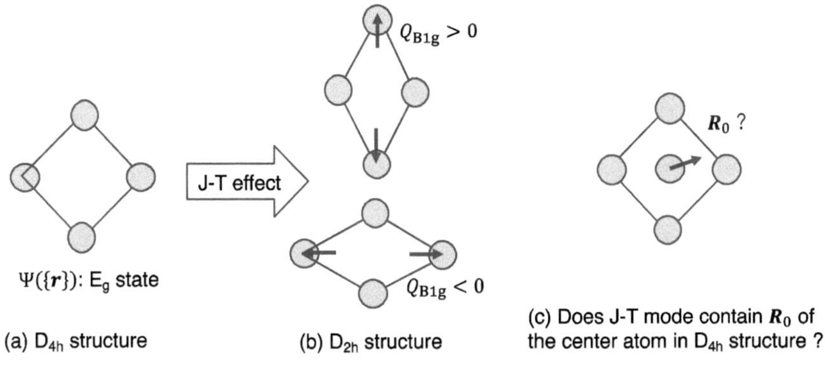

Figure 8.2 An example of the Jahn–Teller effects. (a) D_{4h} structure is unstable for the E_g type electronic states and (b) then distorts to D_{2h}. (c) How about the central atom?

In the rigid lattice $\{\Delta R = 0\}$, $\{Q_\Gamma = 0\}$ all electronic states $\Psi(\{r\})$, which obey the Schrödinger equation

$$[H_e(\{r\}) + V_{eL}(\{r\}, \{Q_\Gamma = 0\})]\Psi(\{r\}) = E\,\Psi(\{r\}), \tag{8.3}$$

are classified by the irreducible representation Γ_e of **G** as $\Psi_{\Gamma_e}(\{\boldsymbol{r}\})$.

Next, we allow atoms to deform. When the electronic state is $\Psi_{\Gamma_e}(\{\boldsymbol{r}\})$, the generalized force $F_{\Gamma_e}^{\Gamma}$ with respect to Q_Γ direction is given by

$$F_{\Gamma_e}^{\Gamma} = \left\langle \Psi_{\Gamma_e} \middle| \frac{\partial V_{eL}}{\partial Q_\Gamma} \middle| \Psi_{\Gamma_e} \right\rangle. \tag{8.4}$$

If $\Psi_{\Gamma_e}(\{\boldsymbol{r}\})$ is degenerate, the symmetric representation $[\Gamma_e \times \Gamma_e]$ always contains at least one non-identity representation Γ. This means that $F_{\Gamma_e}^{\Gamma} \neq 0$ and then atoms always move toward the Q_Γ direction to lower the symmetry: $\bar{Q}_\Gamma \neq 0$. Figure 8.2 shows an example of the J–T distortion at a square type of defect with D_{4h} symmetry. When the electronic state is doubly degenerate E_g, the symmetry is lowered to D_{2h} symmetry. It should be noted that there are several equivalent final stable configurations: In Fig. 8.2 case, there are two equivalent J–T distortions ($\bar{Q}_{B1g} > 0$ and $\bar{Q}_{B1g} < 0$); thus, in the total symmetry remains conserved in statistical sense.

Table 8.1 shows typical examples of the relation between degenerate electronic states and the J–T distortion modes for several point symmetries. A vacancy (V) in the diamond structure has the T_d symmetry. In Si crystal when a vacancy is positively charged V_{Si}^{+}, the electronic ground state is triply degenerate (Γ = T) then the E-type J–T distortion takes place to lower the symmetry to D_{2h}, which has triple equivalent minima. This J–T distortion at V_{Si}^{+} has been observed experimentally by EPR measurements and the potential barrier height between equivalent minimums is estimated as 0.07 eV [11].

So far, we have treated the atomic motion classically. As has been stated in the original paper, the J–T theorem only tells us the initial instability but does not give the magnitude of the atomic distortion and hence that of E_{LR}. When the J–T lattice relaxation energy E_{LR} is smaller than the average phonon energy, $\hbar\bar{\omega}$, quantum transitions occur rapidly among the equivalent minimums of the atomic configuration. If the mean transition time is shorter than the characteristic time of the experiment, the symmetry lowering is zero in average (the dynamic J–T effect). This explains why the J–T effect is practically absent at shallow-level impurities in covalent semiconductors, which have the T_d symmetry in the strict sense. On the other hand, if the electron wave function is localized as is in deep-level defects, electrons interact strongly with the lattice resulting $E_{LR} >> \hbar\bar{\omega}$, then the present J–T instability really takes place and lowers the symmetry for degenerate electronic states.

Table 8.1 Typical examples of the Jahn–Teller distortion modes and the type of the displacement R_0 of the center atom, based on [12]. *Note*: E states are doubly degenerate and T states are triply degenerate

T_d		D_{3d}		C_{3v}	
Electronic state $\Psi_{\Gamma e}$	J–T distortion mode Q_Γ	Electronic state $\Psi_{\Gamma e}$	J–T distortion mode Q_Γ	Electronic state $\Psi_{\Gamma e}$	J–T distortion mode Q_Γ
E	E				
T_1	E, T_2	E_g, E_u	A_{2g}, E_g	E	E
T_2	E, T_2				
R_0 belongs to T_2		R_0 belongs to A_{2u} and E_u		R_0 belongs to A_1 and E	
T_d position is unstable for T_1 and T_2 electronic states		D_{3d} position does not show the instability caused by J–T effect		C_{3v} position is unstable for E electronic state	

8.2 Stability of the Position of an Impurity

Next, let us next apply the above Jahn–Teller theorem to the stability of positions of an impurity atom in semiconductors, either interstitial or substitutional. In the former case, an impurity atom may occupy a position with larger empty space in the crystal, and hence has a high point symmetry. Let us denote the point symmetry around the impurity as **G**. If the electronic state is localized and degenerate, atoms around the impurity are always displaced to lower the symmetry due to the J–T effect as shown above. The question is whether these J–T displacements contain the displacement ΔR_0 of the impurity atom located at the center (Fig. 8.2c). The electronic states usually couple with many atoms around, then the transformation of $\{\Delta R_j\}$ into $\{Q_\Gamma\}$ is laborious. On the other hand, it is much easy to find to which representation $\{Q_\Gamma\}$('s) the central displacement ΔR_0 belongs. The criterion of the instability of the positions of an impurity has been discussed in reference [12] and some of the results are summarized in Table 8.2. Thus, we have useful information on the stability of an impurity position without any elaborate calculations. It should be noted that an impurity atom at any positions which have the inversion symmetry does not show the instability caused by the J–T effect. This is because the irreducible representations for any system with the inversion symmetry are classified into *gerade* and *ungerade*. All the J–T distortion modes are *gerade* while the displacement of the central atom always belongs to *ungerade* modes. Examples are the interstitial hexagonal position (D_{3d}) in the diamond structure, the body-centered position (O_h) and the face-centered position (D_{4h}) in the simple cubic lattice (Table 8.2).

Table 8.2 Defect molecules (a) with and (b) without the inversion symmetry

Symmetry	Inversion symmetry	Degenerated state Γe	J-T distortion Q_Γ		Displacement of central atom
O_h 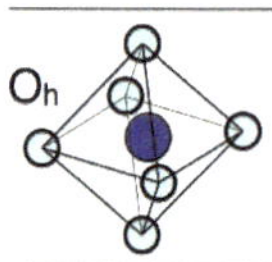	YES	E_g, E_u T_{1g}, T_{1u} T_{2g}, T_{2u}	E_g E_g, T_{2g} E_g, T_{2g}	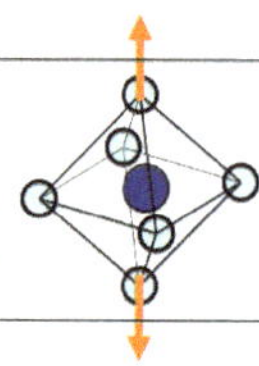	T_{1u} 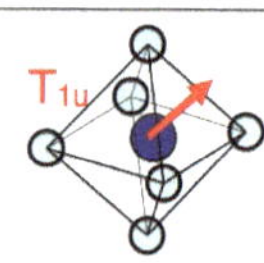
T_d 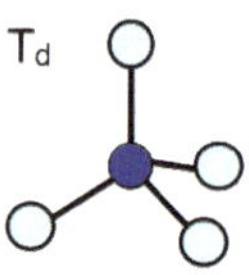	NO	E T_1 T_2	E E, T_2 E, T_2	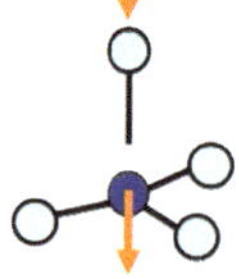	T_2 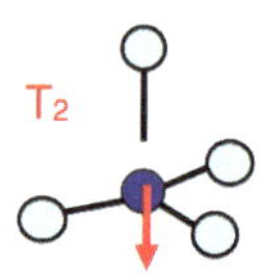

If electronic states $\Psi(\boldsymbol{r})$ explicitly depend on the electron spin, the condition of the Jahn–Teller instability depends on the number of bound electrons. The criterion for even number electrons is whether the symmetric representation $[\Gamma_e \times \Gamma_e]$ contains the displacement ΔR_0 of the impurity, while that for odd number is whether the antisymmetric representation $\{\Gamma_e \times \Gamma_e\}$ contains ΔR_0 or not. The present J–T instability for substitutional impurities corresponds to off-center instability. Examples are N in Si [7, 8], EL2 in GaAs [4], and DX center in AlGaAs [6]. It is surprising that although the T_d is the most popular symmetry met in covalent semiconductors, it shows an inherent instability as shown here. Thus, the diamond and zincblende structures are not so stable as we have been considering with respect to local electronic transitions.

8.3 Bourgoin–Corbett Mechanism

In relation with instability, it is worthy to mention the Bourgoin–Corbett mechanism [13]. Let us consider an interstitial impurity atom in semiconductors. An impurity atom occupies a larger vacant site in the crystal lattice such as **A** and **B** in Fig. 8.3. It is probable that these sites have high point symmetry and then we can apply the results in the previous section. If a localized electronic state which shows the off-center J–T instability at a position **A** which continuously changes into a state which does not show the instability at another position **B**, and vice versa, the diffusion is enhanced by the electronic excitation and also by deexcitaion. This mechanism is similar to the Bourgoin–Corbett (B-C) mechanism [13], but the number of bound electrons (holes) here remains constant. The B-C mechanism has been proposed for an enhancement of an impurity diffusion, where the stable position of an impurity is unstable for another charged state and vice versa (Fig. 8.4). The self-interstitial in Si is found to be this type [14].

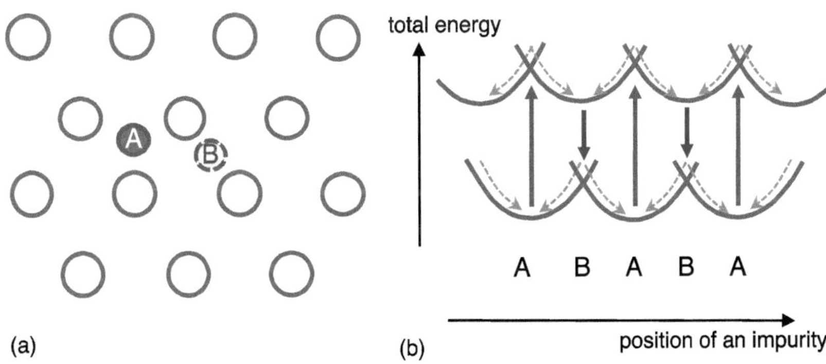

Figure 8.3 (a) Interstitial positions of an impurity atom in a crystal lattice. (b) Electronically enhanced impurity diffusion.

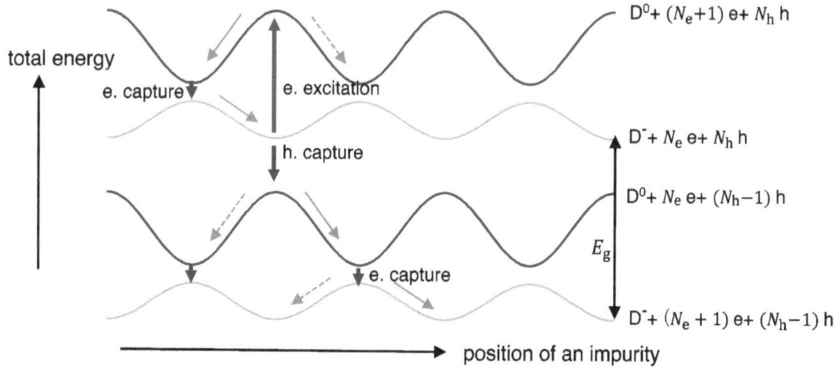

Figure 8.4 The Bourgoin–Corbett mechanism [13] for impurity diffusion.

8.4 Tetrahedral T-U-S Model

In case of strong carrier localization that might happen at deep-level defects in semiconductors, we must take account explicitly of the interaction among localized electrons as well as the electron–lattice interaction (see Appendix VIII). The simplest model to discuss these two effects would be the *T-U-S* model, which consists of several interacting electrons in a defect molecule. The *T-U-S* model has been proposed first by Toyozawa for two-site two-electron model, which is a most conceivable simple model for condensed matter [15]. Extensions to larger systems were done by Schreiber (three-site [16]), Takimoto (four-site [17]), and Shinozuka (three-dimensional bulk [18]). Here we show a study on the tetrahedral four-site *T-U-S* model, which can be applied to (b) a vacancy or (c) a substitutional impurity in the diamond (zincblende) crystal semiconductors [19] (Fig. 8.5).

The Hamiltonian we consider is

$$H = -T \sum_{l,l',\sigma} a_{l,\sigma}^{+} a_{l',\sigma} + U \sum_{l} n_{l,\uparrow} n_{l,\downarrow} - \beta \sum_{l} (n_{l,\uparrow} + n_{l,\downarrow}) Q_l + \frac{1}{2} \sum_{l} Q_l^2 . \tag{8.5}$$

Here, $a_{l,\sigma}^{+}$ ($a_{l,\sigma}$) is the electron creation (annihilation) operator at the l-th site with $\sigma = \uparrow, \downarrow$ spin, and $n_{l,\sigma} \equiv a_{l,\sigma}^{+} a_{l,\sigma}$ is the number operator (l = 1~4). The system is characterized by the electron transfer energy, $-T$, describing the quantum hopping of an electron from site to site, the on-site Coulomb repulsion energy, U, between two electrons with opposite spins at the same site, the lattice relaxation energy, $S(\equiv\beta^2)$, due to the short range interaction of an electron with the lattice mode, Q_l, localized at the l-th site and the number n (= 0~8) of the occupied electrons. The last term is the lattice elastic energy and the origin is chosen to refer as those when electrons are absent (n = 0). The site l represents either (a) atomic or molecular orbital, (b) four dangling bonds for a vacancy and (c) four bonding (or antibonding) orbitals for a substitutional impurity. For case (c), T is equal to the s-p hybridization energy of the impurity atom and Q_l stands for the bond stretching mode between the central impurity and l-th host atom. The electron site energy at l is assumed to change linearly with the lattice distortion Q_l at the site with β the lattice-coupling constant.

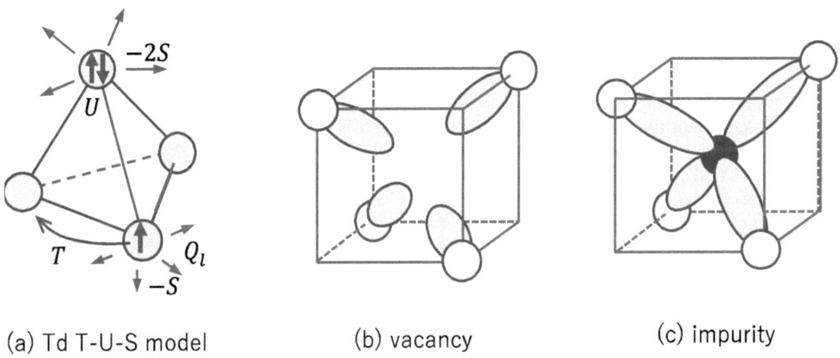

Figure 8.5 (a) T_d *T-U-S* model for (b) vacancy and (c) impurity in tetrahedral semiconductors [19].

The Hamiltonian (8.5) commutes with the number n of electrons and the total spin S_σ, which are then the constants of motion. Diagonalizing the Hamiltonian matrix for each electronic subspace (n, S_σ), one obtains the adiabatic potential $U_i(\{Q_l\})$ both for the ground $\{i = g\}$ and excited states $\{i$ = ex$\}$. The minimum of $U_g(\{Q_l\})$ in four-dimensional $\{Q_l\}$ space corresponds to the stable configuration. Applying the Feynman–Hellmann theorem, the lattice distortion at the stable configuration is shown to be proportional to the local electron density as $\bar{Q}_l = \beta\langle n_{l,\uparrow} + n_{l,\downarrow}\rangle$ and then the sum rule holds, $\sum_l \bar{Q}_l = n\beta$ (this corresponds to the results in Eq. (4.17) in Chapter 4).

In the rigid lattice (Q_l = 0), when the electron–electron interaction is absent (U = 0), one electron eigenstates are one $a1$ state and three $t2$ states. If we allow the lattice to deform ($Q_l \neq 0$) and the electron–electron interaction is switched on ($U \neq 0$), eigenstates of n electrons become complicated and structural changes may take place lowering the symmetry. The stable structures for n (= 0~8) electrons and each spin state are comprehensively studied by El-Maghraby et al. [19] and Takahashi et al. [20]. The stable structures are found to depend on the relative strength among T, U, and S. Some of the results are as follows:

(a) Vacancy

The one-electron energy of a dangling bond is located in the middle of the band gap, and then it is a good approximation to isolate electrons at dangling bonds with other host valence electrons. The lattice distortion here represents a linear combination of the displacements of three host atoms next to the host atom at l (Fig. 8.5b). The vacancy (V) in Si has been found to have different stable structures depending on its charge: D_{2d} for V^0 and V^+, C_{2v} for V^-, and T_d for V^{++}. The neutral V^0 corresponds to n = 4 in the present model, the parameters for V_{Si} may fall into the third row in Table 8.3 (U/T = 1, S/T = 2.4). Since only the diagonal type electron–lattice interaction has been taken into account, the D_{2d} type structure does not appear here. It could be stabilized by the off-diagonal electron–lattice interaction, such as modulation of the electron transfer.

(b) Substitutional impurity

Figure 8.6 shows the phase diagrams of the stable structures for substitutional impurities, displayed in the T-U-S triangle. The coordinates (T, U, S) of any point in this triangle are proportional to the lengths of perpendiculars therefrom to US, ST, TU sides (Fig. 8.6(c)). A singly ionized donor (D^+) and acceptor (A^-) correspond to fully occupied four bonding orbitals and empty four antibonding orbitals. Then both D^+ and A^- correspond to

n = 8 or 0 in the present model. It is found that the T_d structure is found to be always stable there. But when the electronic occupation is changed, a structural instability may take place, which is different for donor and acceptor as can be seen in Fig. 8.6 and Table 8.4. The displacements of the central impurity atom in real space are also illustrated. Neutral D^0 corresponds to n = 1, where fully occupied bonding orbitals have been included into the elastic medium. It is found that T_d structure is stable for $S/T < 6$ and $C_{3v}(1)$ is stable for $S/T > 6$. Nitrogen impurity in Si, which is found to show an off-center configuration for a neutral state [7, 8], may correspond to the latter case. For D^- donor, $C_{3v}(1)$ type off-center configuration is more stable when $S/T > 3$ and $S > U$. DX center in $Al_{1-x}Ga_xAs$ [6] may correspond to this case: T_d structure is stable for neutral donor and $C_{3v}(1)$ off-center structure for DX^- state. Figure 8.7 illustrates the calculated configuration of DX^- by Lany and Zunger using the density functional approximation [21].

Table 8.3 Example of the stable structure of the ground state of T_d *T-U-S* model for different electron occupation n = 0~8 [19]

Parameters	$n = 0$	$n = 1$	$n = 2$	$n = 3$	$n = 4$	$n = 5$	$n = 6$	$n = 7$	$n = 8$
$U/T = 0.4$, $S/T = 0.8$	T_d	T_d	T_d	$C_{3V}(1)$	$C_{3V}(1)$	C	$C_{3V}(2)$	$C_{3V}(2)$	T_d
$U/T = 1$, $S/T = 0.5$	T_d	T_d	T_d	$C_{3V}(1)$	T_d	*T_d	C_{2V}	$C_{3V}(2)$	T_d
$U/T = 1$, $S/T = 2.4$	T_d	T_d	T_d	C_{2V}	C_{2V}	C	$C_{3V}(2)$	$C_{3V}(2)$	T_d
$U/T = 1$, $S/T = 8$	T_d	$C_{3V}(1)$	$C_{3V}(1)$	$C_s(1)$	C_{2V}	C	$C_{3V}(2)$	$C_{3V}(2)$	T_d
$U/T = 4$, $S/T = 2.6$	T_d	T_d	T_d	C_{2V}	T_d	*T_d	C_{2V}	$C_{3V}(2)$	T_d
$U/T = 10$, $S/T = 8$	T_d	$C_{3V}(1)$	C_{2V}	$C_{3V}(2)$	T_d	$C_{3V}(1)$	C_{2V}	$C_{3V}(2)$	T_d
S_σ of the lowest configuration	0	1/2	0	1/2	0	1/2, *3/2	0	1/2	0

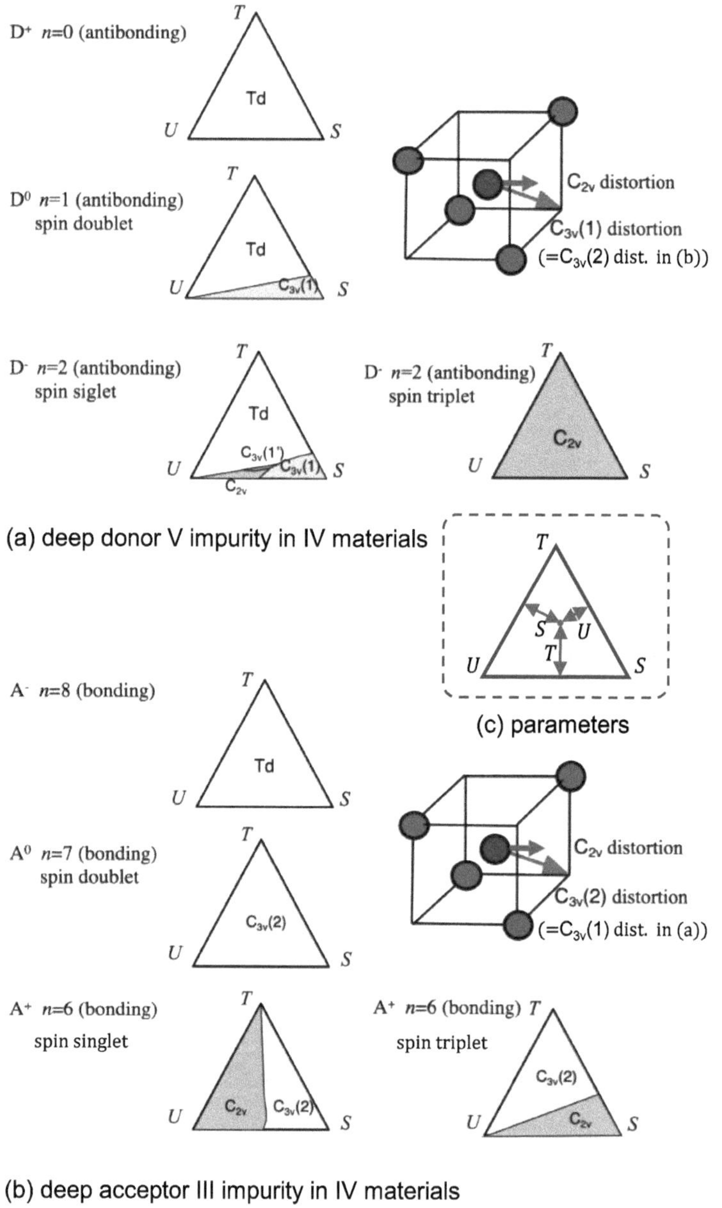

Figure 8.6 Phase diagrams of the ground state structure for a substitutional deep (a) donor or (b) acceptor impurities in semiconductors, (c) Relative strength among *T*, *U*, and *S* shown in the triangle. Reproduced with permission from Ref. [3].

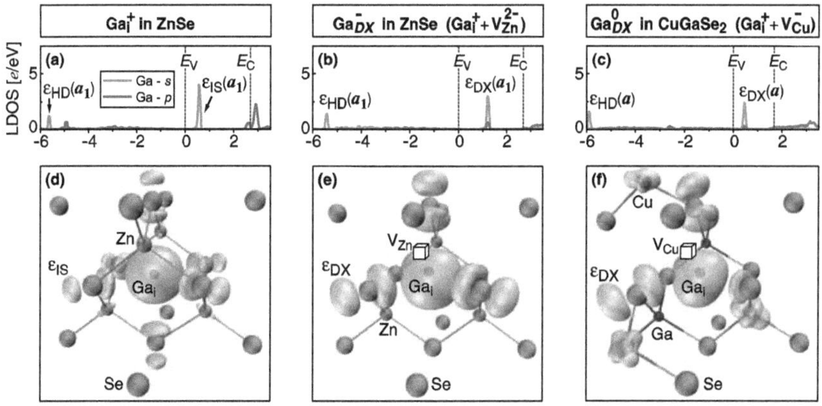

Figure 8.7 Illustration of atomic configuration of DX-center in AlGaAs by first principles calculation. Reproduced with permission from Ref. [21].

Table 8.4 Relation between T_d *T-U-S* model and typical point defects [19]

Vacancy in IV	Charge	V^{2+}	V^+	V^0	V^-	V^{2-}
	n =	2	3	4	5	6
III vacancy in III-V	Charge	V_{III}^{2+}	V_{III}^+	V_{III}^0	V_{III}^-	V_{III}^{2-}
	n =	3	4	5	6	7
V vacancy in III-V	Charge	V_V^{2+}	V_V^+	V_V^0	V_V^-	V_V^{2-}
	n =	1	2	3	4	5
V impurity in IV	Charge	D^{2+}	D^+	D^0	D^-	D^{2+}
	Antibonding		$n = 0$	1	2	3
	Bonding	$n = 7$	$n = 8$			
III impurity in IV	Charge	A^{2+}	A^+	A^0	A^-	A^{2+}
	Antibonding				$n = 0$	1
	Bonding	$n = 5$	6	7	8	

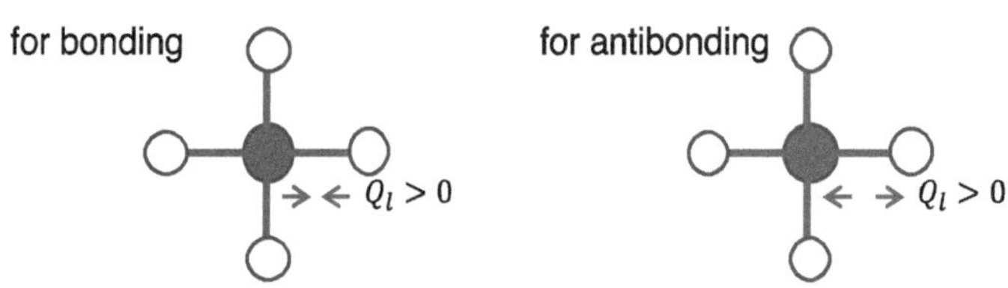

For a neutral acceptor A^0 (corresponds to n = 7) on the other hand, T_d is always unstable and $C_{3v}(2)$ type off-center structure is always stable. Then we can conclude that on-center type neutral deep-acceptor A^0 is absent in tetrahedrally bonded semiconductors. This is not the case for a shallow acceptor A^0, which should be regarded as A^- + one shallowly bound hole in the present approximation. It should be reminded that we have used a crucial approximation where a "defect molecule" has been extracted from the crystal.

It is worthy to conclude that although the tetrahedral coordination is the most popular symmetry met in covalent semiconductors, it is not so rigid as it is considered, especially for hole localization.

(c) Annihilation of an impurity complex

A hydrogen atom can passivate or activate donors, acceptors and other impurities in semiconductors. In silicon, a hydrogen impurity can occupy several stable positions in the diamond lattice. If it occupies a bond center position between a substitutional carbon (isoelectronic) impurity and a silicon atom, this H–C complex acts as a shallow donor whose thermal depth is E_{th}^{e} = 0.15eV. The positively charged H–C complex is stable above the room temperature, while the neutral state is unstable at the room temperature and the activation energy for the annihilation is estimated as 0.5 ~ 0.7eV [22]. Since the "united atom" of an H–C complex is a nitrogen atom in the molecular chemistry, the effective potential seen by a conduction electron away from the H–C complex is similar to that of the N^+ impurity. Then we can derive an analogy between the stability of a N impurity and that of a H–C complex in Si. For a shallow donor state or a positively charged state (an electron is away) the central position is stable for a N impurity, and the H–C complex is stable. When an electron comes close to them, on the other hand, it induces the off-center displacement of a N impurity, and the off-center displacement of a hydrogen atom. That is, a hydrogen atom is pushed away from the carbon: the annihilation of an H–C complex. This would be a typical example of a defect reaction caused by the structural instability mechanism: see Chapter 9.

References

1. Watkins, G. D., *Proc. Defects in Semiconductors*, 1972, 228, *Physica*, **117B & 118B**, 9 (1983).
2. Shinozuka, Y., *Defects in Optoelectronic Materials* ed. by Wada, K., and Pang, S. W. (Gordon and Breach Science Publishers, 2001) Chapter 8.
3. Shinozuka, Y., *Materials and Reliability Handbook for Semiconductor Optical and Electron Devices*, ed. by Ueda, O., and Pearton, S. (Springer, 2012) Chapter 10.
4. Chadi, D. J., and Chang, K. J., *Phys. Rev. Lett.*, **60** (1988) 2187.
5. Hida, A., Mera, Y., and Maeda, K., *Physica B: Condens. Matter*, **308–310** (2001) 738.
6. Dabrowiski, J., and Scheffler, M., *Phys. Rev. Lett.*, **60** (1988) 2183.
7. Brower, K. L., *Phys. Rev. B*, **26** (1982) 6040.
8. Murakami, K., Kuribayashi, H., and Masuda, K., *Phys. Rev. B*, **38** (1988) 1589.
9. Jahn, H. A., and Teller, E., *Proc. Roy. Soc.*, **A161** (1937) 220.
10. Onodera, Y., *Group Theory and Application to Quantum Mechanics* (Springer, 1995).
11. Sprenger, M., Muller, S., Sieverts, E., and Ammerlaan, C., *Phys. Rev. B*, **35** (1987) 1566.
12. Shinozuka, Y., *Materials Science Forum*, **83–87** (1992) 527.
13. Bourgoin, J. C., and Corbett, J. M., *Phys. Lett.*, **38A**, 135 (1972); and *Radiation Effects*, **36** (1978) 157.
14. Saito, M., Oshiyama, A., and Sugino, O., *Phys. Rev. B*, **47** (1993) 13205.
15. Toyozawa, Y., *J. Phys. Soc. Jpn.*, **51** (1981) 1861.
16. Schreiber, M., *J. Phys. Soc. Jpn.*, **56** (1987) 1029.
17. Takimito, J., *J. Phys. Soc. Jpn.*, **67** (1998) 3524.
18. Shinozuka, Y., *J. Phys. Soc. Jpn.*, **56** (1987) 4477.
19. El-Maghraby, M., and Shinozuka, Y., *J. Phys. Soc. Jpn.*, **67** (1998) 3524.
20. Takahashi, I., and Shinozuka, Y., *Physica B*, **340–342** (2003) 349.
21. Lany, S., and Zunger, A., *Phys. Rev. Lett.*, **100** (2008) 016401.
22. Kamiura, Y., Tsutsune, M., Hayashi, M., Yamashita, Y., and Hashimoto, F., *Mater. Sci. Forum*, **196–201** (1995) 903.

Chapter 9

Modification of Semiconductors: Destruction and Creation

We have studied in this book the origin of the electron–lattice interaction in semiconductors and various induced effects. Strong electron–lattice interaction in semiconductors causes carrier localization, lattice distortion, energy dissipation and relaxation, structural change, etc. In this chapter, we summarize our journey in semiconductor physics and further explore possibilities of novel atomic process induced by electronic excitation with attention to modification of the structure and search for new structures.

9.1 Return to the Beginning: Condensed Matter

In the beginning of this book, Chapter 1, we started with a basic Hamiltonian of condensed matter, which consists of a huge number of ions and valence electrons. In the ground state, a homogeneous material shows a crystal structure, where component atoms (ions) are spatially arrayed in periodic. The corresponding translational symmetry governs the electronic structures, representing the Bloch theorem. In non-metallic materials, there is a fundamental energy gap dividing the electron states into the valence and

Electron–Lattice Interactions in Semiconductors
Yuzo Shinozuka

ISBN 978-981-4800-96-9 (Hardcover), 978-1-003-02346-3 (eBook)
www.jennystanford.com

conduction bands. The presence of the band gap guarantees the stiffness of the group of the valence electrons and hence the elastic lattice. Across the band gap, various physical phenomena take place between carriers, the lattice vibrations (phonons) and radiation fields (photons), as have seen in this book.

The electron–lattice interaction is originated from the imbalance in the interatomic force mediated by valence electrons generated by electronic excitation, showing the difference of the stable positions in the electronic ground and excited states in semiconductors. When the interaction between a carrier (conduction electron, valence hole) and the lattice (phonon) is weak, the effects can be treated as a weak perturbation: An example is a decrease in the carrier mobility. As the interaction strength increases, a carrier self-trapping takes place resulting in strong localization, which is usually seen in ionic materials, and also in covalent semiconductors with the assistance of an imperfection: extrinsic self-trapping. Strong electron–lattice interaction causes large atomic displacements in nonmetallic materials, such as bistability and off-center instability. Once a carrier is localized, an electronic energy order of ~eV is converted to the lattice by emitting phonons and stored in the elastic energy. The induced atomic displacements are limited in *local*, which inevitably destroys the crystal symmetry: a local imperfection, i.e., a new "defect" is created tentatively or permanently [1].

9.2 Strong Electron–Lattice Interaction: Defect Process

Once electrons in a semiconductor are excited optically or by carrier injection, this will induce atomic rearrangements, whose degree depends on materials and sometimes generates a local structural change, i.e., defect reaction. Figure 9.1 schematically shows typical examples of defect reactions [2–4]: impurity diffusion, creation and annihilation of defects, and climb and glide motions of a dislocation, etc. Hereafter we indicate defect reactions symbolically as $D_i \rightarrow D_f$. Several models for the mechanisms have been proposed so far, which can be classified into two categories:

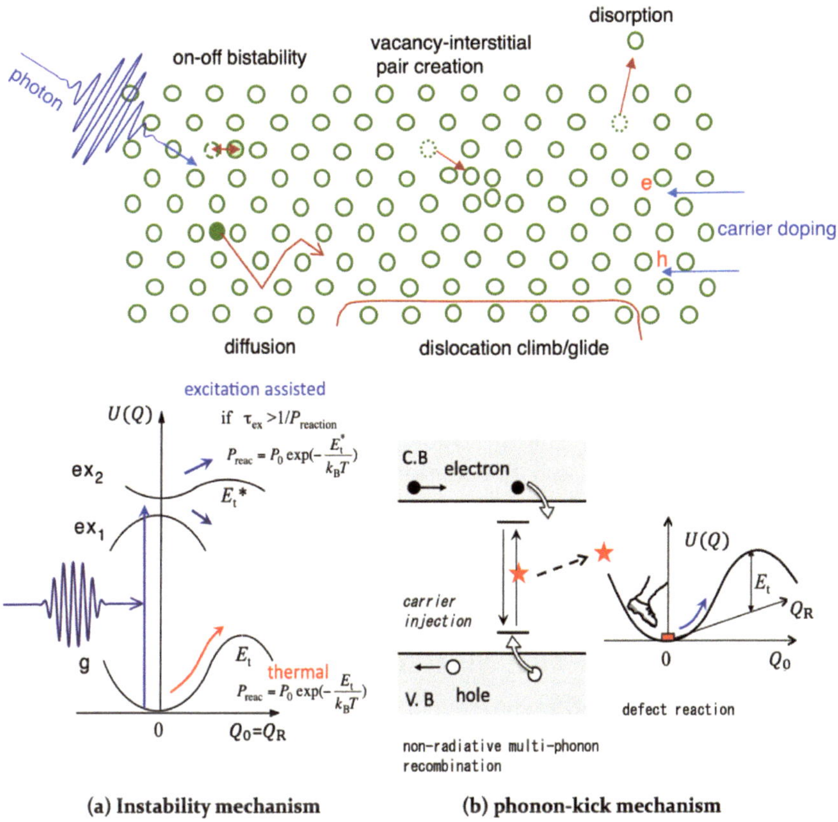

Figure 9.1 Electronic excitations induce various defect reactions in semiconductors. Reproduced with permission from Ref. [3].

(a) Instability mechanism

One is the structural instability mechanism [4], where a particular electronic state of a defect induces a symmetry breaking atomic motion such as the off-center type distortion. For example, if the relaxation mode Q_0 is nothing but the reaction coordinate Q_R ($D_i \rightarrow D_f$) itself, an electronic excitation to ex1 promptly induces the reaction or generates a precursor of D_f. In another case, an excitation to ex2 decreases the potential barrier height ($E_t \rightarrow E_t^*$) of the reaction. The activation energy is then decreased in the excited state whose lifetime τ is much larger than the inverse of the reaction rate ($1/P_{reac}$). The velocity of b-dislocation in p-GaAs is enhanced under illumination, and then the mechanism is considered to be this type [5]. The Bourgoin–Corbett mechanism and the electronically enhanced impurity diffusion are classified into this category (see Chapter 8).

(b) Phonon-kick mechanism

The other mechanism is the recombination enhanced (or phonon-kick) mechanism, where nonradiative multiphonon recombination(s) of carriers at a defect induces a transient atomic motion in the reaction coordinate Q_R [4, 6].

The first mechanism (a) stresses why and how the instability takes place toward a "defect" configuration. The second mechanism (b) stresses how the electronic energy converts to the lattice vibration energy and the motion in the reaction coordinate to overcome the potential barrier. Although two mechanisms look different, they can be treated in a unified scheme. First, let us have a look at the phonon-kick mechanism.

9.3 Phonon-Kick Mechanism for Defect Reactions

In Chapter 7, we discussed the lattice vibrations at defects within a harmonic approximation. If both carrier densities are high enough to make coherent successive carrier captures possible, the amplitude of lattice vibrations may increase beyond the harmonic region. In this section, we try to analyze the possibility of defect creation by the phonon-kick mechanism.

(1) If the relaxation mode Q_0 does not break the point symmetry of the defect such as the breathing mode, Q_0 is then orthogonal to Q_R. Thus, the induced lattice vibrations caused by carrier captures do not affect defect reactions.

(2) If the reaction coordinate Q_R is almost parallel to Q_0, the phonon-kick mechanism takes place with the aid of $E_{act}^{e} + E_{th}^{e}$ induced by electron capture: a defect reaction whose potential barrier height is E_t takes place through a single capture in nonzero regions in Fig. 7.3(a), if we replace the abscissa $E_{act}^{h}/(E_{act}^{e} + E_{th}^{e})$ by $E_t/(E_{act}^{e} + E_{th}^{e})$. In the zero

region, the phonon-kick mechanism needs an inflation of the lattice vibration caused by a series of multiple carrier captures.

(3) If $Q_0 // Q_R$, the phonon-kick mechanism may work through a hole capture that has been enhanced by an electron capture. The band gap energy $E_g = E_{th}^{e} + E_{th}^{h}$ can be used to excite the defect reaction. This could be the most probable path to the defect reaction, which is nothing but the so-called recombination-enhanced mechanism.

(4) When Q_0 is not orthogonal to Q_R, Sumi [7, 8] showed theoretically that the transient lattice vibration due to a capture reduces the activation energy E_t for a defect reaction to

$$E_t - \left(E_{act}^{i} + E_{th}^{i}\right) \text{ or } \left[\sqrt{E_t} - \cos\theta\sqrt{E_{act}^{i} + E_{th}^{i}}\right]^2 / (1 - \cos^2\theta), \quad (9.1)$$

where θ is the angle between Q_0 and Q_R (Fig. 9.2).

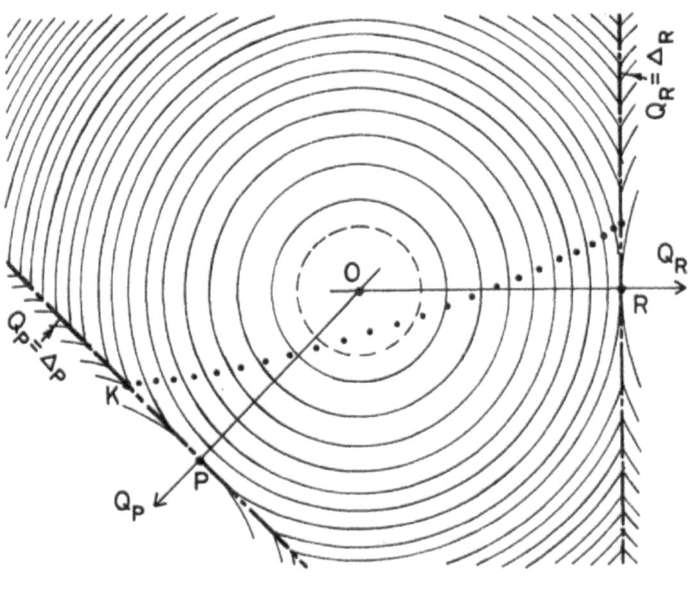

Figure 9.2 (a) Thermal activation energy E_t ($\equiv E_H$) for the defect reaction as a function of the phonon-kick energy $E_{act}^{i} + E_{th}^{i}$ ($\equiv E_P$), where $g = \cos\theta$. (b) Adiabatic potential surface shown in a plane of the interaction mode Q_0 ($\equiv Q_P$) and the reaction coordinate Q_R. After Sumi [7].

(5) In case of off-center type distortions, which is found at some deep-level defects, the symmetry in Q_0 and $-Q_0$ does not exist. In that case, it is important to assign which corresponds to the off-center configuration, $Q_0 = 0$ or $\bar{Q}_0$.

(6) In Section 7.1, we have assumed that the defect has the D^0 and D^- states. If it has D^0 and D^+, the present discussion can also be valid with the respective translations of $D^0 \rightarrow D^+$ and $D^- \rightarrow D^0$. As have shown, either in n-type and p-type semiconductors, a majority carrier capture is always enhanced by a minority carrier capture. If the Coulomb attraction is considered, an electron (hole) capture by D^+ (D^-) is also accelerated.

9.4 Defect Processes

Let us remember the off-center instability introduced in Section 8.3, which takes place by a charge change of the defect. It is important which corresponds to the off-center configuration: $Q_0 = 0$ or $\bar{Q}_1$? When the reaction coordinate Q_R is almost parallel to Q_1, defect reactions in p-type semiconductors occurs by the structural instability mechanism induced by a minority electron capture (Fig. 9.3). If E_{th}^{e} is large and $Q_0//Q_R$ but there is a potential barrier, the phonon-kick mechanism takes place with the aid of E_{th}^{e} induced by an electron capture. If E_{th}^{h} is large and $Q_0//-Q_R$, the phonon-kick mechanism takes place by a hole capture which has been enhanced by an electron capture. If a capture of a majority carrier occurs very quickly within the local transient vibration, the band gap energy $E_g = E_{th}^{e} + E_{th}^{h}$ can be used for an excitation of the defect reaction. This could be the most effective path to the defect reaction and nothing but the so-called recombination enhanced mechanism.

In n-type semiconductors, defect reactions can be discussed in a similar way as above, starting from $Q_0 = \bar{Q}_0$ and D_i^0+e+h. In contrast to p-type, the first capture of a minority hole by a neutral center D_i^0 is not accelerated by the Coulomb attraction, while the second electron capture by D_i^+ is accelerated. Up to here we have assumed that the defect has D_i^0 and D_i^+ states. If it has D_i^- and D_i^0, the above discussion can be also valid with respective translations with $D_i^0 \rightarrow D_i^-$ and $D_i^+ \rightarrow D_i^0$. As we have shown, either in n-type and p-type, the capture of a majority carrier always enhanced by a minority carrier capture, and vice versa. But only a hole capture is accelerated by the Coulomb attraction of D_i^-.

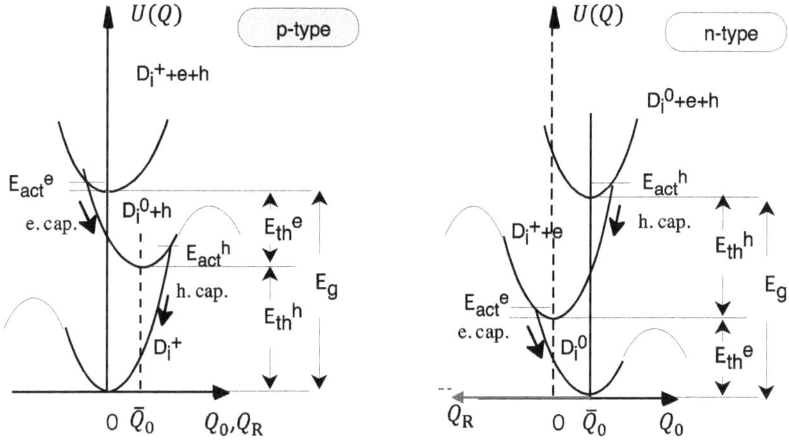

	capture	defect reaction: $Q_0//Q_R$	defect reaction: $Q_0//-Q_R$
p-type	electron thermally activation	⇒ struct. instability if $E_{th}^e \sim E_g$ phonon kick	
p-type	hole enhanced	⇒	if $E_{th}^h \sim E_g$ recomb. enhanced
n-type	hole thermally activation	⇒	struct. instability if $E_{th}^h \sim E_g$ phonon kick
n-type	electron enhanced	⇒ if $E_{th}^e \sim E_g$ recomb. enhanced	

Figure 9.3 A typical configuration coordinate diagram for the same defect in p-type and n-type conditions. The table summarizes the most probable path for defect reactions in relation with the interaction mode Q_0 (relaxation mode) and the reaction coordinate Q_R. Reproduced with permission from Ref. [3].

9.5 Control the Electron–Lattice Interaction

We have learned that the electron–lattice interaction is closely related to condensation of material and is realized as a dynamical response of condensed matter when it is excited [9]. In this section, we would like to discuss how to control this inherent interaction.

(1) The coupling constant of the electron–lattice interaction is specific to material itself; then it is difficult to control. Alloying is an effective technique to change the strength of the interaction. As seen in Section 4.4, the manifestation of self-trapping can be changed by alloying: the F–S change of a polaron in silver halides and extrinsic self-trapping in thallium halides.

(2) One possible way to weaken the effect is decrease the phonon population, which is realized by lowering the temperature. Since the available phonon number is decreased at low temperature, the carrier scattering rate is decreased, hence increase of the carrier mobility. See Section 2.3. However, the phonon emission process still remains even at low temperature.

(3) In Chapter 4, we have seen that extrinsic self-trapping of a carrier and exciton is induced by crystal imperfections, the decrease in dimension, and confinement in space. So, a suitable design of quantum well structure would control the manifest of self-trapping.

(4) Another possible way to decrease the effect would be harden or soften the lattice, such as stiffness parameter. If a semiconductor material is fabricated on a substrate of a hard material, the effect of the electron–lattice interaction in the semiconductor will be decreased. The reverse can be also expected.

(5) A direct method to quit the interaction is to stop the presence of electronic excitation: no carrier nor exciton. In relation to the absence, we can investigate the following attractive situations.

9.6 Expolar Novel Metastructures in the Dark

At the absolute zero temperature, the stable atomic arrangement of a material is determined by the minimum of the total energy of all electrons and ions. Usually, a bulk of uniform material chooses one of the crystal structures with a certain translational symmetry (Section 1.3). The chosen structure is confirmed to be realized only when it is stable with respect to thermal excitations

($\sim k_B T$). But this is not enough. It should be noted that there is an additional stability condition for realization of the structure. Under a normal environment on earth (our planet), every material is exposed to various external radiation, mostly coming from the sun. Then realized material structures should be stable against any photoexcitations across the electron band gap (~eV). This second condition is not usually mentioned explicitly but could be very essential and worthy to be examined.

How about if this second condition is excluded? That is, is it possible for those material structures to exist—stable with respect to thermal excitations but not to photoexcitations across the electron band gap. This is a basic question in materials science. The following phenomena have been found in relation to this conjecture (Fig. 9.4).

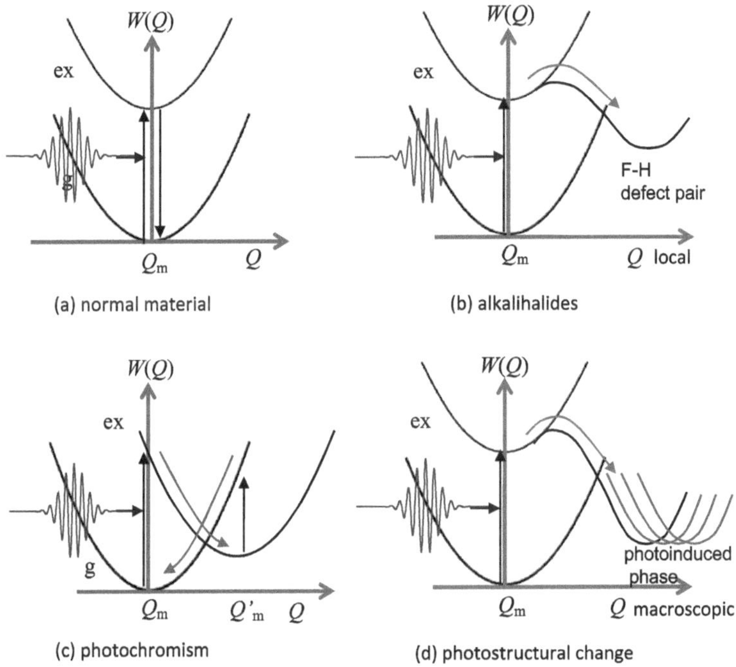

Figure 9.4 Possible structural changes in various materials illustrated in configuration coordinate diagrams.

(1) Photograph exposure phenomenon: When silver halides (AgCl, AgBr) are irradiated by visible light, structural change

occurs to create silver clusters (latent image), which later gather with each other to create gigantic clusters after developing.

(2) Defect generation in alkali halides by photo irradiation: When alkali halides (NaCl, KCl, NaBr, etc.) are irradiated by photons across the band gap (3–5 eV), halogen ions move in (110) direction to form a self-trapped exciton, and an F–H defect pair is generated (see Fig. 4.18). The created F–H defect pair becomes extinct immediately at room temperature [10, 11]. It would be fascinating if we imagine a periodic array of many "F–H defect pairs" in alkali halides, which could be realized by intense photoexcitations.

(3) Photochromism: Reversible transformation of a chemical species between two forms by irradiation, where the two forms have different absorption spectra [12]. As for azobenzene, for example, cis → trans by UV light irradiation; isomerizes it and the molecular head shortens. It changes into a transformer when irradiated by visible light and return to original length afterwards.

(4) Photo-induced phase transition: For example, TTF-CA shows the first-order phase transition thermally between low-temperature ionic phase and high-temperature neutral phase [13–16]. When it is irradiated by light at low temperature a domain of metastable neutral phase is transiently generated (Fig. 9.5). The Domino mechanism has been proposed to explain the experimental results. A new phase can be fabricated also by cooperative multiple photoexcitations.

(5) Light evocation solid-phase polymerization and light induced phase transition of diacetylene: The single crystal of 112-hexadiyne diol screw phenyl carbamate (TCDU) takes up promptly polymerization by light at room temperature, and it is with a polydiacetylene PDA, but does not polymerize it by heating [17]. Furthermore, a kind of PDA-4U3 of polydiacetylene shows reversible light induced phase transition.

These are natural outcrops of "metastable" materials, which make us to expect a possibility of novel metastable material structures with respect to thermal excitation but not for optical excitations. They could be obtained if we try to fabricate materials in darkness. Complete darkness is not necessary, but such a cave is enough for the purpose which shut photons above visible region (Fig. 9.6). That is, a cave is similar to the darkroom of the photograph developing. These new category of metastable materials, "Gollum material,"[1] could open a new area of materials science and technology.

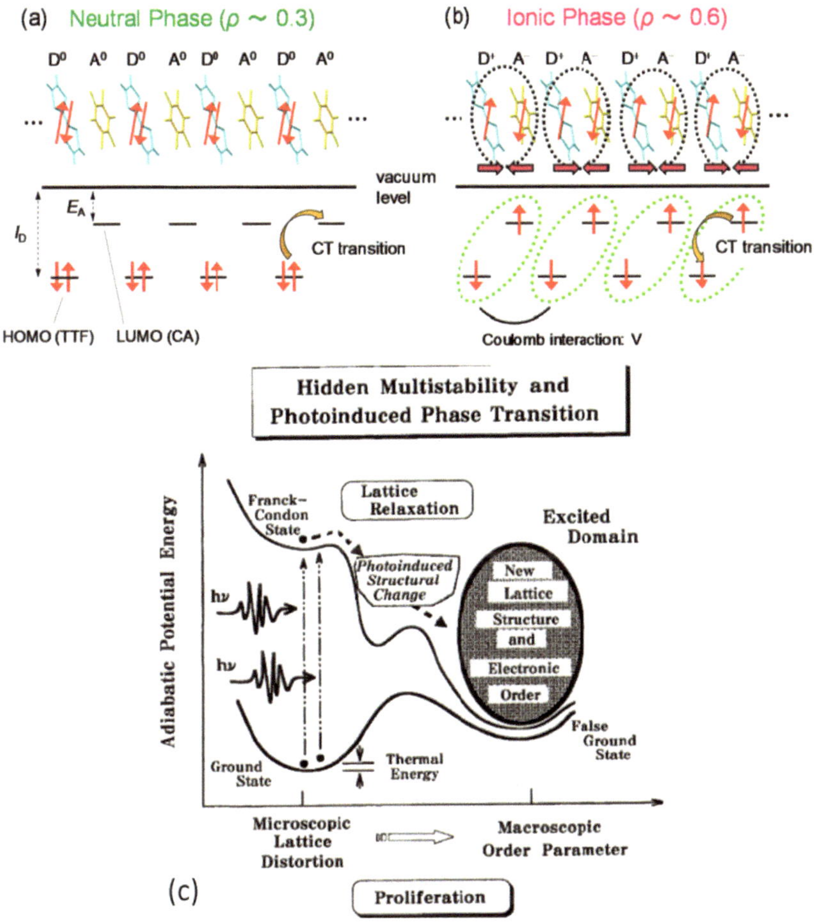

Figure 9.5 (a) Illustration of electronic structures in (a) the neutral phase and (b) the ionic phase of TTF-CA. Reproduced from Ref. [16]. (c) Physical image of photo-induced phase transition. Reproduced with permission from Ref. [15].

[1]After a character "Gollum" in a book of "The Lord of the Ring" by J. R. R. Tolkin.

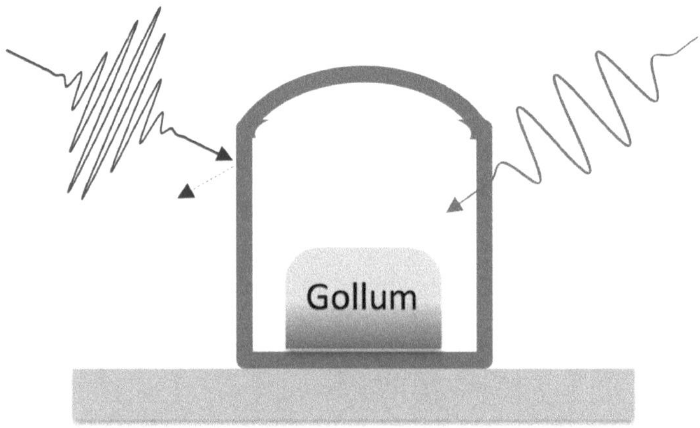

Figure 9.6 A novel "Gollum material" can be fabricated in a "dark" cave.

Table 9.1 Relations between normal and Gollum materials

		Gollum material		
Stability	**Ordinal material**	**Photochromism**	**Photo-induced phase transition**	**Novel**
1st condition stable against thermal excitation	○	○	○	○
2nd condition stable against optical excitation	○	×	×	×
In relaxed excited states	$Q_m = Q_{m'}$	Bistability	Cooperative	Instability defect creation destruction

References

1. Itoh, N., and Stonham, A. M., *Materials Modification by Electronic Excitation* (Cambridge Univ. Press, Cambridge, 2001).
2. Ueda, O., *Reliability and Degradation of III-V Optical Devices* (Artech House Pub., Boston-London, 1996).
3. Shinozuka, Y., in *Materials and Reliability Handbook for Semiconductor Optical and Electron Devices*, ed. by Ueda, O., and Pearton, S. (Springer-Verlag, Berlin, 2012), Chapter 10.

4. Shinozuka, Y., *Proc. MRS 1996 Fall Meeting*, **442**; *Defects in Electronic Materials II*, ed. Michel, J., Kennedy, T., Wada, K., and Thonke, K., p. 225.
5. Maeda, K., and Takeuchi, S., in *Dislocations in Solids*, ed. by Nabarro, F. R. N., and Duesbery, M. S. (North-Holland, Amsterdam, 1996), vol. 10, pp. 444–504.
6. Shinozuka, Y., *Mater. Res. Soc. Symp. Proc.*, Vol. 1195 B02-02 (2009).
7. Sumi, H., *Phys. Rev. B*, **29** (1984) 4616.
8. Sumi, H., *J. Phys. C*, **17** (1984) 6071.
9. Toyozawa, Y., *Optical Processes in Solids* (Cambridge Univ. Press, Cambridge, 2003).
10. Kan'no, K., Tanaka, K., and Hayashi, T., *Rev. Solid State Sci.*, **4** (1990) 383.
11. Song, K. S., and Williams, R. T., *Self-Trapped Excitons* (Springer-Verlag, Berlin, 1993).
12. Irie, M., *Chem. Rev.*, **100** (2000) 1685.
13. Nasu, K. (ed.), *Relaxations of Excited States and Photo-Induced Structural Phase Transitions* (Springer-Verlag, Berlin, 1997).
14. Nasu, K., *Photo-Induced Phase Transitions* (World Scientific, Singapore, 2004).
15. Nasu, K., *Eur. Phys. J. B*, **75** (2010) 415.
16. Morimoto, T., Miyamoto, T., and Okamoto, H., *Crystals*, **7**(5) (2017) 132.
17. Itoh, C., Kondoh, T., and Tanimura, K., *J. Phys. Soc. Jpn.*, **68** (1999) 1711.

Appendix I

Adiabatic Approximation

Consider a simple situation where an electron (r, p) couples with the atomic motion (R, P) via interaction $V_{\mathrm{eI}}(r, R)$. The total Hamiltonian is

$$H = \left[\frac{p^2}{2m} + V_{\mathrm{e}}(r)\right] + V_{\mathrm{eI}}(r, R) + \left[\frac{P^2}{2M} + V_{\mathrm{I}}(R)\right]$$

$$\equiv H_{\mathrm{e}} + V_{\mathrm{eI}}(r, R) + H_{\mathrm{I}}. \tag{A1.1}$$

Here m is the electron mass and M the atomic (ionic) mass. Since $M >> m$, the adiabatic approximation can be applied as follows. For a fixed atomic coordinate, R, we first solve the electron part

$$[H_{\mathrm{e}} + V_{\mathrm{eI}}(r, R)]\psi_{R,l}(r) = E_l(R)\psi_{R,l}(r). \tag{A1.2}$$

Here $E_l(R)$ and $\psi_{R,l}(r)$ are the l-th eigenenergy and eigenstate of the electron, each contains the variable R as a parameter. Next, introducing the adiabatic potential

$$U_l(R) = E_l(R) + V_{\mathrm{I}}(R), \tag{A1.3}$$

the atomic motion is considered to be obeyed through the following equation:

$$\left[\frac{P^2}{2M} + U_l(R)\right]\chi_{l,n}(R) = E_{l,n}\chi_{l,n}(R). \tag{A1.4}$$

Thus, the wave function of electron-atom is assumed to be given by

$$\Phi_{l,n}(r, R) = \psi_{R,l}(r)\chi_{l,n}(R), \tag{A1.5}$$

where l denotes the electronic level and n stand for the atomic level under fixed l (Fig. A1.1).

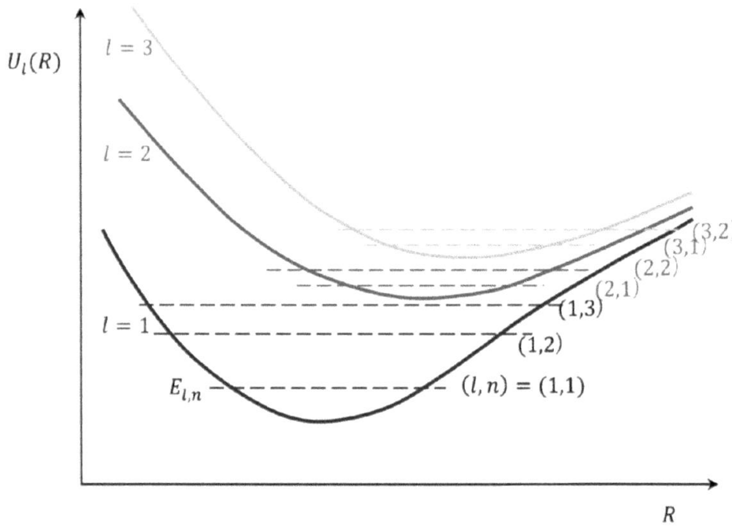

Figure A1.1 Thick solid line indicates the adiabatic potential $U_l(R)$ in R-space and horizontal dotted line indicates the energy level $E_{l,n}$ of the adiabatic state $\psi_{R,l}(r)\chi_{l,n}(R)$ where l stands for electron state and n for atom under fixed l.

However, this is not the true eigenstates of the total Hamiltonian, i.e.,

$$H\,\psi_{R,l}(r)\chi_{l,n}(R) \neq E_{l,n}\psi_{R,l}(r)\,\chi_{l,n}(R). \tag{A1.6}$$

This can be seen as follows. Define the adiabatic Hamiltonian

$$H_{\text{adia}} \equiv \frac{p^2}{2m} + V_{\text{e}}(r) + V_{\text{eI}}(r, R) + V_{\text{I}}(R), \tag{A1.7a}$$

$$\text{i.e.,} \quad H = H_{\text{adia}} + \frac{P^2}{2M}. \tag{A1.7b}$$

Since H_{adia} operates only to electron part $\psi_{R,l}(r)$, we can have

$$H_{\text{adia}}\psi_{R,l}(r) = [E_n(R) + V_{\text{I}}(R)]\,\psi_{R,l}(r) = U_l(R)\,\psi_{R,l}(r). \tag{A1.8}$$

Then

$$\begin{aligned} H\Phi_{l,n}(r,R) &= \left[\frac{P^2}{2M} + H_{\text{adia}}\right]\psi_{R,l}(r)\chi_{l,n}(R) \\ &= \left[\frac{P^2}{2M} + U_l(R)\right]\psi_{R,l}(r)\chi_{l,n}(R) \\ &= \chi_{l,n}(R)\left[\frac{P^2}{2M} + U_l(R)\right]\psi_{R,l}(r) + E_{l,n}\,\psi_{R,l}(r)\chi_{l,n}(R). \end{aligned} \tag{A1.9}$$

The quantity $\chi_{l,n}(R)\left[\frac{P^2}{2M} + U_l(R)\right]\psi_{R,l}(r)$ does not vanish, because $P = \frac{\hbar}{i}\frac{\partial}{\partial R}$ and $\psi_{R,l}(r)$ contains the atomic coordinate R as a parameter.

For further analysis beyond the adiabatic approximation, let be the perturbation H' defined by

$$H'\,\psi_{R,l}(r)\chi_{l,n}(R) = \chi_{l,n}(R)\left[\frac{P^2}{2M} + U_l(R)\right]\psi_{R,l}(r) \tag{A1.10}$$

This perturbation H' will generate a transition to different adiabatic states $\psi_{R,l'}(r)\chi_{l',n'}(R)$ and also mix with $\psi_{R,l}(r)\chi_{l,n'}(R)$ and $\psi_{R,l'}(r)\chi_{l',n'}(R)$. These transition and mixing would be probable when $E_{l,n} \approx E_{l',n'}$. Otherwise, the adiabatic approximation is appropriate.

Appendix II

Einstein Relation between Photoabsorption and Photoemission Spectra

Consider a system where optical transitions take place between a group g of ground states $\{i\}$ and a group ex of excited states $\{j\}$, well separated by an energy order of 1eV, much larger than the thermal energy $k_B T$ (k_B is the Boltzmann constant and T is the absolute temperature) (Fig. A2.1). Within the same group, the eigenenergy E_i and E_j distribute closely. A typical example is a molecule or a defect in semiconductors, where two electronic states g and ex are coupled with phonon states to create vibronic states distributed with a distance of the phonon energy ~meV. We assume that the optical transitions occur between a group of ground states $\{i\}$ and a group of excited states $\{j\}$.

(a) Photoabsorption

Before the absorption, the thermal equilibrium has been achieved in the electronic ground states $\{i\}$ with the Boltzmann distribution

$$f_g(E_i) = \frac{1}{Z_g} g_i \exp[-\beta E_i], \qquad Z_g \equiv \sum_{i \in g} g_i \exp[-\beta E_i], \quad \text{(A2.1)}$$

Here g_i is the degeneracy in the i-th state, $\beta = 1/k_B T$, and Z_g a normalization factor. The absorption spectrum with photon energy $h\nu$ is given by

$$I_{ab}(h\nu) = \sum_{i \in g} \sum_{j \in ex} f_g(E_i)\, B_{ij}\, \delta(E_j - E_i - h\nu). \quad \text{(A2.2)}$$

Here B_{ij} is the transition probability $i \to j$, which is historically called the Einstein B coefficient, and is basically given by the Fermi's golden rule. Delta function represents the energy conservation.

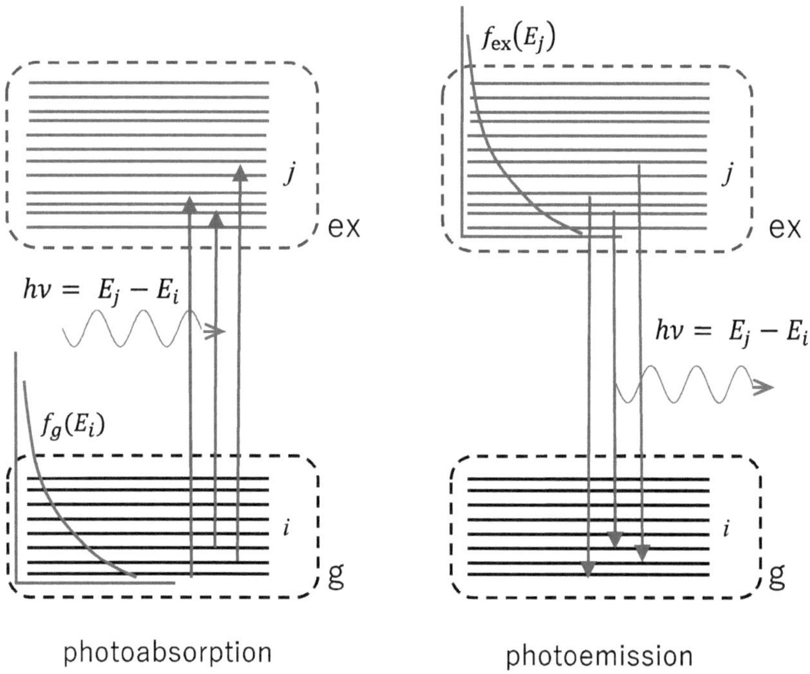

Figure A2.1 Optical transitions between a group of ground states $\{i\}$ and a group of excited states $\{j\}$.

(b) Photoemission

Before photoemission, if (quasi) thermal equilibrium has achieved within the electronic excited states $\{j\}$, the distribution is given by

$$f_{ex}(E_j) = \frac{1}{Z_{ex}} g_j \exp[-\beta E_j], \qquad Z_{ex} \equiv \sum_{j \in ex} g_j \exp[-\beta E_j] \quad \text{(A2.3)}$$

Here g_j is the degeneracy in the j-th state. The (spontaneous) emission spectrum with a photon energy $h\nu$ is given by

$$I_{em}(h\nu) = \sum_{j \in ex} \sum_{i \in g} f_{ex}(E_j)\, A_{ji}\, \delta(E_j - E_i - h\nu). \quad \text{(A2.4)}$$

Here A_{ji} is the transition probability $j \to i$, which is historically called the Einstein A coefficient. Here we recall the relation between Einstein's A coefficient and B coefficient,

$$A_{ji} = B_{ij}\frac{4hv^3}{c^3}, \tag{A2.5}$$

and also, the detailed balance between $j \leftrightarrow i$: $g_j B_{ji} = g_i B_{ij}$. Then

$$I_{\rm em}(hv) = \sum_{j\in \rm ex}\sum_{i\in \rm g}\frac{1}{Z_{\rm ex}} g_j \exp[-\beta E_j]\ A_{ji}\ \delta(E_j - E_i - hv)$$

$$= \sum_{j\in \rm ex}\sum_{i\in \rm g}\frac{1}{Z_{\rm ex}} g_j \exp[-\beta E_j]\ B_{ij}\frac{4hv^3}{c^3}\ \delta(E_j - E_i - hv)$$

$$= \sum_{j\in \rm ex}\sum_{i\in \rm g}\frac{1}{Z_{\rm ex}} g_i \exp[-\beta(E_i + hv)]\ B_{ji}\frac{4hv^3}{c^3}\ \delta(E_j - E_i - hv)$$

$$= \frac{Z_{\rm g}}{Z_{\rm ex}}\frac{4hv^3}{c^3}\exp[-\beta hv]\sum_{i\in \rm g}\sum_{j\in \rm ex} f_{\rm g}(E_i)\ B_{ji}\ \delta(E_j - E_i - hv).$$

Thus, we obtain a general relation between the photo absorption and photoemission spectra:

$$I_{\rm em}(hv) = \frac{Z_{\rm g}}{Z_{\rm ex}}\frac{4hv^3}{c^3}\exp[-\beta hv]\, I_{\rm ab}(hv), \qquad \beta = \frac{1}{k_B T}. \tag{A2.6}$$

Appendix III

Relation between Phonon States and Shifted Phonon States

Let us consider a harmonic vibration system of the ordinate q with mass M and angular frequency ω (Fig. A3.1). When the equilibrium position is $q = 0$, the phonon eigenstates are given by

$$\Phi^{\mathrm{L}}_{\mathrm{g},n}(q) = N_n H_n(\alpha q)\exp\left[-\frac{1}{2}\alpha^2 q^2\right], \quad n = 0,1,2,\ldots. \tag{A3.1}$$

Here $H_n(z)$ is the n-th Hermite polynomial, $\alpha \equiv \sqrt{M\omega/\hbar}$, and the normalization constant $N_n = \pi^{-1/4}\sqrt{\alpha/2^n n!}$. When the equilibrium position is $q = \bar{q}$, the phonon eigenstates are

$$\Phi^{\mathrm{L}}_{\mathrm{ex},n'}(q') = N_{n'} H_{n'}(\alpha q')\exp\left[-\frac{1}{2}\alpha^2 q'^2\right], n' = 0,1,2,\ldots. \tag{A3.2}$$

Here $q' \equiv q - \bar{q}$, and the lattice relaxation energy is $E_{\mathrm{LR}} = \frac{1}{2}M\omega^2\bar{q}^2$ in comparison with two ground states $\Phi^{\mathrm{L}}_{\mathrm{g},0}(q)$ and $\Phi^{\mathrm{L}}_{\mathrm{ex},0}(q')$.

The relation between $\Phi^{\mathrm{L}}_{\mathrm{g},n}(q)$ and $\Phi^{\mathrm{L}}_{\mathrm{ex},n'}(q')$ can be easily obtained with use of the relation

$$\Phi^{\mathrm{L}}_{\mathrm{g},n}(q) = \Phi^{\mathrm{L}}_{\mathrm{ex},n}(q') = \Phi^{\mathrm{L}}_{\mathrm{ex},n}(q - \bar{q}). \tag{A3.3}$$

Taking a Taylor expansion of the right-hand side [A3.1]

$$\Phi^{\mathrm{L}}_{\mathrm{g},n}(q) = \Phi^{\mathrm{L}}_{\mathrm{ex},n}(q) + (-\bar{q})\frac{d}{dq}\Phi^{\mathrm{L}}_{\mathrm{ex},n}(q) + \frac{1}{2!}(-\bar{q})^2\frac{d^2}{dq^2}\Phi^{\mathrm{L}}_{\mathrm{ex},n}(q) + \cdots$$

$$= \exp\left[-\bar{q}\frac{d}{dq}\right]\Phi^{\mathrm{L}}_{\mathrm{ex},n}(q)$$

$$= \exp\left[\bar{q}\sqrt{\alpha/2}(b^+ - b)\right] \Phi^{\mathrm{L}}_{\mathrm{ex},n}(q)$$

$$= \exp[-\alpha^2\bar{q}^2/4] \exp\left[\bar{q}\sqrt{\alpha/2}\, b^+\right] \exp\left[-\bar{q}\sqrt{\alpha/2}\, b\right] \Phi^{\mathrm{L}}_{\mathrm{ex},n}(q). \tag{A3.4}$$

Here, b^+ and b are the creation and annihilate operator defined by

$$b^+ = \sqrt{\frac{\omega}{2\hbar}}\left(q - \frac{i}{\omega}p\right), \quad b = \sqrt{\frac{\omega}{2\hbar}}\left(q + \frac{i}{\omega}p\right), \quad p = \frac{\hbar}{i}\frac{\partial}{\partial q}. \tag{A3.5}$$

The relation (A3.4) holds for any phonon state n.

In special, for the zero-phonon state, $n = 0$, we have a relation

$$\Phi^{\mathrm{L}}_{\mathrm{g},0}(q) = \exp[-\alpha^2\bar{q}^2/4] \exp\left[\bar{q}\sqrt{\alpha/2}\, b^+\right] \Phi^{\mathrm{L}}_{\mathrm{ex},0}(q)$$

$$= \sum_n \frac{\left(\sqrt{\alpha/2}\,\bar{q}\right)^n}{\sqrt{n!}} \Phi^{\mathrm{L}}_{\mathrm{ex},n}(q). \tag{A3.6}$$

In addition, we obtain

$$\left|\langle \Phi^{\mathrm{L}}_{\mathrm{ex},n} | \Phi^{\mathrm{L}}_{\mathrm{g},0} \rangle\right|^2 = \frac{S^n}{n!}\exp[-S]. \tag{A3.7}$$

or inversely,

$$\left|\langle \Phi^{\mathrm{L}}_{\mathrm{g},n} | \Phi^{\mathrm{L}}_{\mathrm{ex},0} \rangle\right|^2 = \frac{S^n}{n!}\exp[-S], \tag{A3.8}$$

Here

$$S \equiv \frac{E_{\mathrm{LR}}}{\hbar\omega} = \frac{M\omega\bar{q}^2}{2\hbar}.$$

is the Huang-Rhys factor, see section 3.5.

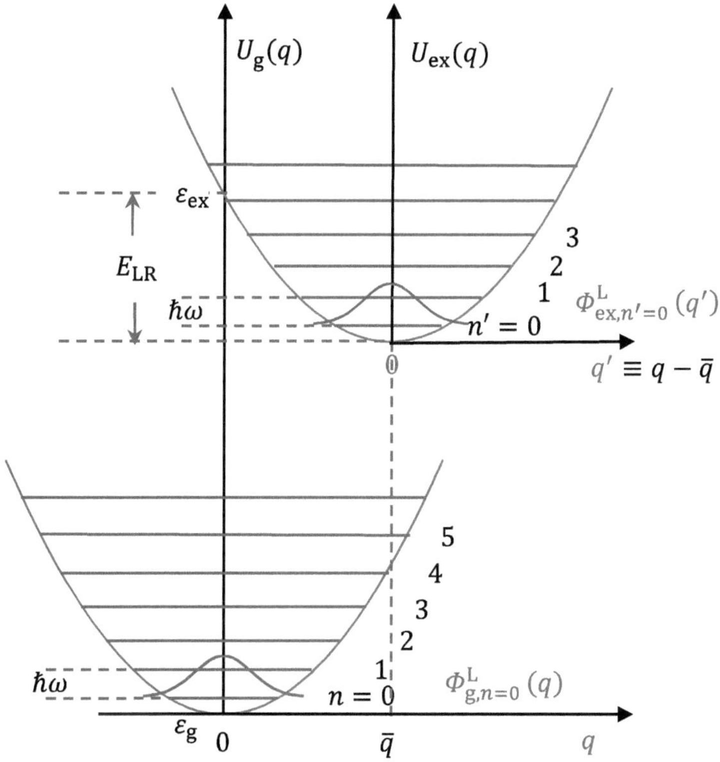

Figure A3.1 Relation between the phonon states $\Phi^L_{g,n}(q)$ and the shifted phonon states $\Phi^L_{ex,n'}(q')$.

Reference

[A3.1] Haken, H., *Quantum Field Theory of Solids*, North-Holland, Amsterdam, 1983.

Appendix IV

Landau–Zener–Stückelberg Formula in Level Crossing

Consider an electronic transition in a two-level system coupled with a time dependent classical variable Q. The Hamiltonian is

$$H = |\psi_1\rangle\varepsilon_1(Q)\langle\psi_1| + |\psi_2\rangle\varepsilon_2(Q)\langle\psi_2| + |\psi_1\rangle b\langle\psi_2| + |\psi_2\rangle b\langle\psi_1|. \tag{A4.1}$$

Here, b is a transfer (transition) energy between two electronic states $|\psi_1\rangle$ and $|\psi_2\rangle$. The diagonal energy are assumed to depend on Q as

$$\varepsilon_1 = \frac{1}{2}\delta - \frac{1}{2}Q, \quad \varepsilon_2 = -\frac{1}{2}\delta + \frac{1}{2}Q, \tag{A4.2}$$

illustrated by solid line in Fig. A4.1. The eigen energy of (A4.1), which corresponds the adiabatic potential (except the elastic energy) is

$$U_{\mathrm{up,low}}(Q) = \frac{1}{2}\left[(\varepsilon_1 + \varepsilon_2) \pm \sqrt{(\varepsilon_1 - \varepsilon_2)^2 + 4b^2}\right], \tag{A4.3}$$

illustrated by dotted line in Fig. A4.1.

In the adiabatic approximation, the time evolution of $Q(t)$ is obeyed by the Newton equation similar to Eq. (1.29) with adding the remaining potential to $U_{\mathrm{up,low}}(Q)$. But here, Q is considered to be an external variable which varies with time as $Q(t) = vt$. Then the energy difference varies as $\varepsilon_1(t) - \varepsilon_2(t) = \delta - vt$.

The time-dependent Schrödinger equation is

$$i\hbar\frac{\partial}{\partial t}|\psi(t)\rangle = H\,\psi(t), \qquad |\psi(t)\rangle = c_1(t)|\psi_1\rangle + c_2(t)|\psi_2\rangle. \tag{A4.4}$$

Starting with the initial condition $c_1(0) = 1$ and $c_1(0) = 0$, we want to obtain $|\psi(t >> t_c)\rangle$, where $t_c = \delta/v$ is the time of crossing. Thus, $|c_2(t >> t_c)|^2$ expresses the transition probability from $|\psi_1\rangle$ to $|\psi_2\rangle$. Landau, Zener, and Stückelberg showed that the probability is approximately given by

$$|c_1(\infty)|^2 = \exp\left[-\frac{2\pi b^2}{\hbar v}\right], \tag{A4.5a}$$

$$|c_2(\infty)|^2 = 1 - \exp\left[-\frac{2\pi b^2}{\hbar v}\right]. \tag{A4.5b}$$

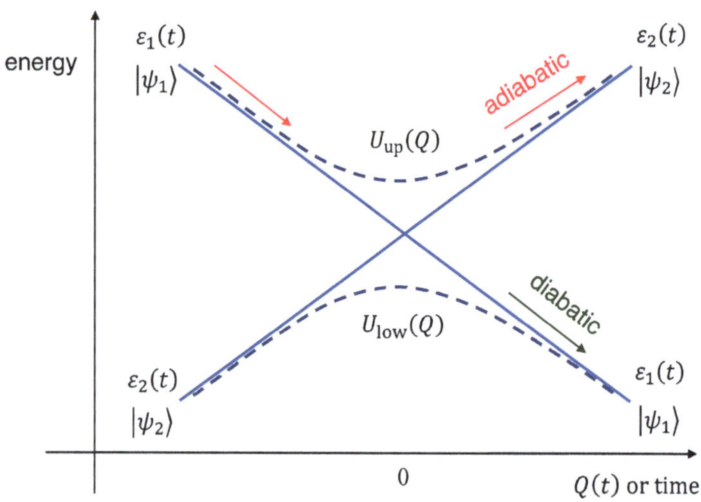

Figure A4.1 Landau–Zener–Stückelberg level crossing problem.

The details of calculation are found in Refs. [A4.1–A4.3]. Figure A4.2 shows results of numerical calculation, directly solving Eq. (A4.4). The formula (A4.5a,b) tells us that

$$\text{If } 2\pi b^2/\hbar v \ll 1, \quad |c_2(\infty)|^2 = 0 \text{ diabatic: fast process} \tag{A4.6a}$$

$$\text{If } 2\pi b^2/\hbar v \gg 1, \quad |c_2(\infty)|^2 = 1 \text{ adiabatic: slow process} \tag{A4.6b}$$

The Landau-Zener-Stückelberg formula can be applicable to various transition problems in electron–phonon coupled system.

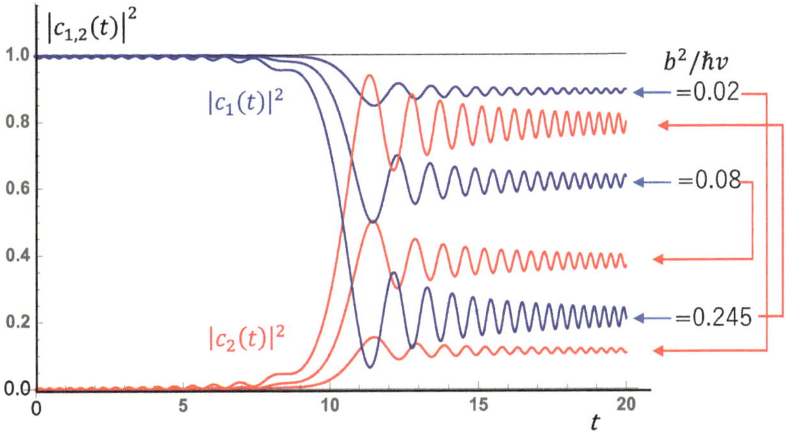

Figure A4.2 Results of numerical simulations on the level-crossing with parameters $c_1(0) = 1$, $|\varepsilon_1(0) - \varepsilon_2(0)| = 10$, and $v = 10$. After the crossing at $t_c = 10$. the probabilities $|c_1(t)|^2$ and $|c_2(t)|^2$ tend to values, indicated by the horizontal position of each arrow, obtained by the Landau–Zener-Stückelberg formulas (A4.5a) and (A4.5b), respectively.

References

[A4.1] Landau, L., *Physikalische Zeitschrift der Sowjetunion.* **2** 46–51 (1932).

[A4.2] Zener, C., *Proc. Royal Soc. Lond. A.* **137** (6) 696–702 (1932).

[A4.3] Stueckelberg, E. C. G., *Helvetica Phys. Acta.* **5**, 369 (1932).

Appendix V

Tight-Binding Model for Electronic States in a Cubic Crystal

It is informative to investigate electronic states in crystals using a simple tight binding model. Consider a crystal composed of a rectangular array of atoms (Fig. A5.1), each located at

$$\boldsymbol{T}_l = l_1\boldsymbol{a}_1 + l_2\boldsymbol{a}_2 + l_3\boldsymbol{a}_3, \quad l_1, l_2, l_3\text{: integer,} \tag{A5.1}$$

$$|\boldsymbol{a}_1| = |\boldsymbol{a}_2| = |\boldsymbol{a}_3| = a.$$

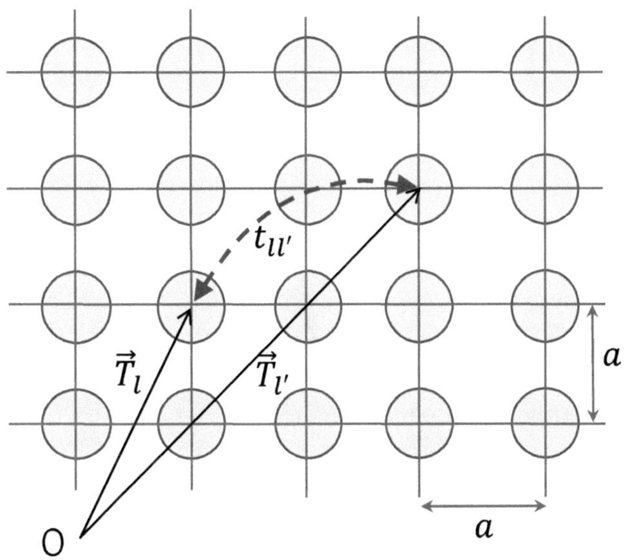

Figure A5.1 A tight binding model: an electron moves in a simple cubic crystal via transfer energy $t_{ll'}$.

Each atom is accompanied by an atomic orbital $\phi_{T_l}(\boldsymbol{r})$. Hereafter we use bra and ket presentation as $\langle\phi_{T_l}|$ and $|\phi_{T_{l'}}\rangle$, in order to simplify the expression.

$$\langle\phi_{T_l}|\phi_{T_{l'}}\rangle = \delta_{ll'} \qquad \text{orthogonality} \tag{A5.2}$$

$$\sum_l |\phi_{T_l}\rangle\langle\phi_{T_l}| = 1. \qquad \text{completeness} \tag{A5.3}$$

Hamiltonian of an electron is expressed as

$$H_0 = \sum_l |\phi_{T_l}\rangle\, \varepsilon\, \langle\phi_{T_l}| + \sum_l \sum_{l'} |\phi_{T_l}\rangle\, t_{ll'} \langle\phi_{T_{l'}}|, \tag{A5.4}$$

Here ε is the site diagonal (atomic) energy (not depend on l) and $t_{ll'}$ is the transfer energy between $\langle\phi_{T_l}|$ and $|\phi_{T_{l'}}\rangle$. The suffix 0 in H_0 denotes perfect crystal. Due to the translational symmetry of the crystal, $t_{ll'}$ is a function of the relative distance $\boldsymbol{T}_l - \boldsymbol{T}_{l'}$. The eigenstates of Hamiltonian (A5.4) are Bloch type with k-vector given by Eq. (1.13b),

$$|\psi_{\boldsymbol{k}}\rangle = (\sqrt{N})^{-1} \sum_l \exp(i\boldsymbol{k}\cdot\boldsymbol{T}_l)\, |\phi_{T_l}\rangle, \tag{A5.5a}$$

or inversely, $$|\phi_{T_l}\rangle = (\sqrt{N})^{-1} \sum_k \exp(-i\boldsymbol{k}\cdot\boldsymbol{T}_l)\, |\psi_{\boldsymbol{k}}\rangle, \tag{A5.5b}$$

The corresponding eigenenergy is,

$$\varepsilon_{\boldsymbol{k}} = \varepsilon + N^{-1} \sum_l \sum_{l'} \exp[i\boldsymbol{k}\cdot(\boldsymbol{T}_l - \boldsymbol{T}_{l'})]\, t_{ll'}. \tag{A5.5c}$$

Here N is the number of total sites. The energy density of states per site is

$$\rho_0(E) = N^{-1} \sum_{\boldsymbol{k}} \delta(E - \varepsilon_{\boldsymbol{k}}). \tag{A5.6}$$

Lattice Green's Function

The lattice Green's operator $G(z)$ and function $F(z)$ would be very useful quantities, defined by

$$G(z) \equiv \frac{1}{z-H},$$

Wannier base $F_{ll'}(z) \equiv \langle \phi_{T_l} | G(z) | \phi_{T_{l'}} \rangle = \langle \phi_{T_l} | \frac{1}{z-H} | \phi_{T_{l'}} \rangle$, (A5.7)

Bloch base $F_{kk'}(z) \equiv \langle \psi_k | G(z) | \psi_{k'} \rangle = \langle \psi_k | \frac{1}{z-H} | \psi_{k'} \rangle$. (A5.8)

When the system has a translational symmetry, i.e. Hamiltonian is given by such as (A5.4), we have

$$G_0(z) \equiv \frac{1}{z-H_0},$$

$$F_{0\,kk'}(z) \equiv \langle \psi_k | \frac{1}{z-H_0} | \psi_{k'} \rangle$$
$$= \langle \psi_k | \frac{1}{z-\varepsilon_{k'}} \left| \psi_{k'} \right\rangle = \frac{1}{z-\varepsilon_k} \delta_{kk'}. \qquad \text{(A5.9)}$$

$$F_{0\,ll'}(z) \equiv \langle \phi_{T_l} | \frac{1}{z-H_0} | \phi_{T_{l'}} \rangle$$

$$= N^{-1} \sum_k \sum_{k'} \exp(i\boldsymbol{k}' \cdot \boldsymbol{T}_l) \exp(-i\boldsymbol{k} \cdot \boldsymbol{T}_{l'}) \langle \psi_k | \frac{1}{z-H_0} | \psi_{k'} \rangle.$$

$$= N^{-1} \sum_k \frac{\exp[i\boldsymbol{k} \cdot (\boldsymbol{T}_l - \boldsymbol{T}_{l'})]}{z-\varepsilon_k}. \qquad \text{(A5.10)}$$

The pole of the Green's function (i.e., the denominator is equal to zero) corresponds to the eigenenergy of the system: ε_k in crystal case.

Let us recall a mathematical useful identity

$$\frac{1}{x - i\,0_+} = \mathrm{P}\frac{1}{x} + i\,\pi\,\delta(x). \tag{A5.11}$$

Here, P denotes an operator which states in case of integration it should be done at principal value, $\delta(x)$ is Dirac's delta function, and 0_+ is a positive infinitesimal small value. After the definition in Eq. (A5.6), the energy density of states is shown to be related with the Green's function as

$$\rho(E) = -\pi^{-1}\mathrm{Im}[F_{0\,ll}(E - i0_+)]. \tag{A5.12}$$

$$F_{0\,ll}(z) = N^{-1}\sum_{\boldsymbol{k}} \frac{1}{z - \varepsilon_{\boldsymbol{k}}}. \tag{A5.13}$$

Additional note on a general Hamiltonian

$$H = \sum_{l} |\phi_l\rangle\,\varepsilon_l\langle\phi_l| + \sum_{l}\sum_{l'} |\phi_l\rangle\,t_{ll'}\langle\phi_{l'}|, \tag{A5.14}$$

Here, the site energy ε_l and the transfer energy $t_{ll'}$ can have arbitral values. On the eigenenergy E_i of the Hamiltonian (A5.14) (the suffix i is not restricted to the wave vector but arbitral), there holds a relation,

$$E_{\mathrm{sum}} = \sum_{i} E_i = \mathrm{Tr}\,H = \sum_{l} \varepsilon_l. \tag{A5.15}$$

Here Tr represents a summation of all diagonal elements in the matrix. The relation of Eq. (A5.15) is valid for any system with arbitral values of diagonal and nondiagonal elements.

Nearest neighbor transfer model

When the transfer energy is limited to only between the nearest neighbor of $\langle\phi_{T_l}|$ and $|\phi_{T_{l'}}\rangle$, which is equal to $-V(<0)$, the eigenenergy is given by

$$\varepsilon_{\boldsymbol{k}} = \varepsilon - 2\,V\big[\cos(k_x a) + \cos(k_y a) + \cos(k_z a)\big] \tag{A5.16}$$

Figure A5.2 shows the density of states for one-dimensional, two-dimensional, and three-dimensional rectangular crystals.

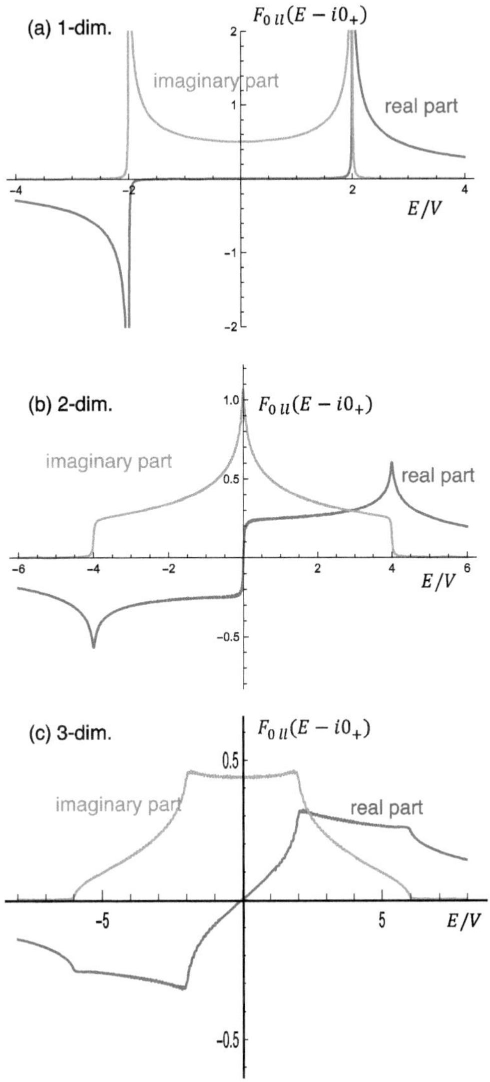

Figure A5.2 The lattice Green's function ($F_{0ll}(z)$ in Eq. (A5.12)) for (a) one-dimensional, (b) two-dimensional, and (3) three-dimensional rectangular crystals with the nearest neighbor transfer $-V(<0)$. The imaginary part presents the density of states $\rho(E)$. Here we choose $\varepsilon = 0$, and $0_+ = 0.01\,V$ in Eq. (A5.12).

Appendix VI

Study of Short Range Defect Potential with Green's Function Method

Consider the same tight binding model introduced in Appendix V with addition of a point defect (impurity or vacancy) located at the origin (Fig. A6.1). Hamiltonian of an electron is expressed as

$$H = |\phi_0\rangle\Delta\langle\phi_0| + \sum_l |\phi_l\rangle\, \varepsilon\langle\phi_l| + \sum_l \sum_{l'} |\phi_l\rangle\, t_{ll'}\langle\phi_{l'}|,$$
$$\equiv H' + H_0, \tag{A6.1}$$

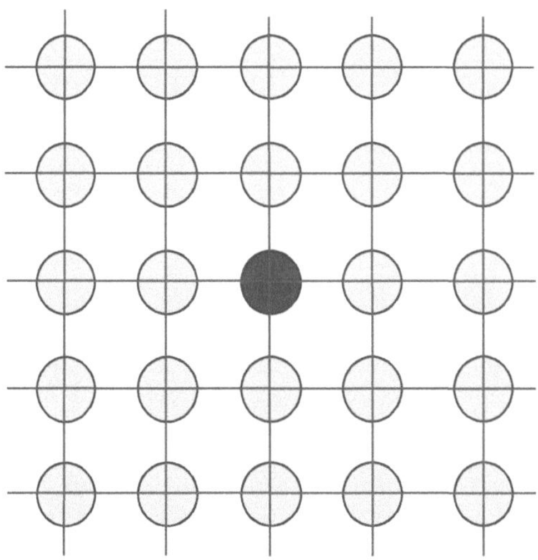

Figure A6.1 A point defect located at the origin (red circle).

The quantity Δ denotes the strength of the short range potential at the origin felt by an electron: $\Delta > 0$ corresponds to repulsive and $\Delta < 0$ corresponds to attractive potential. A vacancy can be treated by assuming $\Delta = \pm\infty$.

Let us define Green's operator as

$$G(z) \equiv \frac{1}{z - H} \quad \text{or} \quad G^{-1}(z) = z - H. \tag{A6.2}$$

Using a Taylor expansion with respect to H', we have a relation

$$G(z) \equiv \frac{1}{z - H_0 - H'} = \frac{1}{G_0^{-1}(z) - H'}$$
$$= G_0(z) + G_0(z)H'G_0(z) + G_0(z)H'G_0(z)H'G_0(z) + \cdots \tag{A6.3}$$

The corresponding Green's function is then

$$F_{00}(z) \equiv \langle\phi_0|G(z)|\phi_0\rangle$$
$$= \langle\phi_0|G_0(z)|\phi_0\rangle + \langle\phi_0|G_0(z)H'G_0(z)|\phi_0\rangle$$
$$+\langle\phi_0|G_0(z)H'G_0(z)H'G_0(z)|\phi_0\rangle + \cdots . \tag{A6.4}$$

Insert the identity operator $\sum_l|\phi_l\rangle\langle\phi_l| = 1$ into every between H' and $G_0(z)$. Since non zero matrix elements are only $\langle\phi_0|H'|\phi_0\rangle = \Delta$, we find

$$F_{00}(z) = F_{0\,00}(z) + F_{0\,00}(z)\,\Delta\,F_{0\,00}(z) + F_{0\,00}(z)\,\Delta\,F_{0\,00}(z)\Delta\,F_{0\,00}(z) + \cdots$$

$$= \frac{1}{F_{0\,00}^{-1}(z) - \Delta}, \qquad \text{with } F_{0\,00}(z) \equiv \langle\phi_0|G_0(z)|\phi_0\rangle. \tag{A6.5}$$

Then, the eigenenergy E of the defect system $H = H' + H_0$ is given by its pole, i.e.,

$$F_{0\,00}^{-1}(E) - \Delta = 0 \quad \text{or} \quad \frac{1}{\Delta} = F_{0\,00}(E) = \frac{1}{N}\sum_k \frac{1}{E - \varepsilon_k}. \tag{A6.6}$$

Figure A6.2 schematically shows the relation between eigenenergy $\varepsilon_i(i = k)$ of H_0 (crystal) and E_i of H (with defect). It can be readily seen that

$$\varepsilon_{i-1} < E_i < \varepsilon_i \quad \text{for} \quad \Delta < 0 \text{ case.} \tag{A6.7}$$

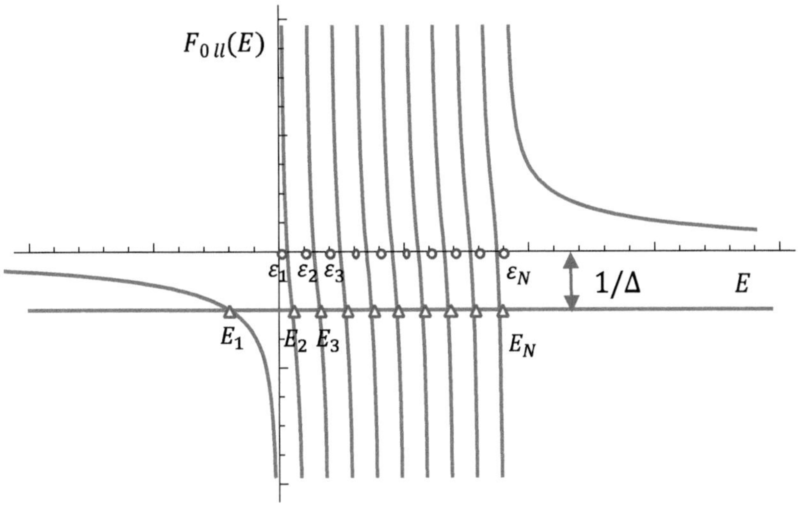

Figure A6.2 Relation between the eigenenergies ε_l (small circle) of H_0 and E_l (small triangle) of H.

Almost all solutions do not change remarkably as $E_l \sim \varepsilon_l$, except at the top or below of the body band. In $N \to \infty$ limit the spectrum of E_l is the same as that of ε_l of the host band. If the potential is attractive: $\Delta < 0$ (repulsive: $\Delta > 0$) a localized level may split from the body of the band and is located below (above).

Nearest neighbor transfer model

When the transfer energies are limited to between nearest neighbors of $|\phi_l\rangle$ and $|\phi_{l'}\rangle$, which is equal to $-V$ (usually negative), the lattice Green's function in the crystal is given by

$$F_{0\,ll}(z) = \frac{1}{N}\sum_{k_x}\sum_{k_y}\sum_{k_z} \frac{1}{z - \{\varepsilon - 2V[\cos(k_x a) + \cos(k_y a) + \cos(k_z a)]\}}$$

$$= \frac{1}{(2\pi)^3}\int_0^{2\pi}\int_0^{2\pi}\int_0^{2\pi} \frac{1}{z - \{\varepsilon - 2V[\cos x + \cos y + \cos z]\}} dxdydz. \tag{A6.8}$$

In one dimension, the integration can be done analytically, and the result is

$$\mathrm{Re}F_{0\,ll}(E) = \begin{cases} \dfrac{1}{\sqrt{(E-\varepsilon)^2 - 4\,V^2}} & E > \varepsilon + 2V \\ 0 & \varepsilon - 2V < E < \varepsilon + 2V \\ \dfrac{-1}{\sqrt{(E-\varepsilon)^2 - 4\,V^2}} & E < \varepsilon - 2V \end{cases} \quad \text{(A6.9a)}$$

$$\mathrm{Im}F_{0\,ll}(E) = \begin{cases} \dfrac{1}{\sqrt{4\,V^2 - (E-\varepsilon)^2}} & \varepsilon - 2V < E < \varepsilon + 2V \\ 0 & \text{otherwise} \end{cases}. \quad \text{(A6.9b)}$$

As far as $\Delta \neq 0$, a localized state appears whose level is given by

$$E(\Delta) = \varepsilon + \sqrt{\Delta^2 + 4\,V^2} \quad (\Delta > 0: \ \text{repulsive}) \quad \text{(A6.10a)}$$

$$E(\Delta) = \varepsilon - \sqrt{\Delta^2 + 4\,V^2} \quad (\Delta < 0: \ \text{attractive}). \quad \text{(A6.10b)}$$

The results are shown in Fig. A6.3.

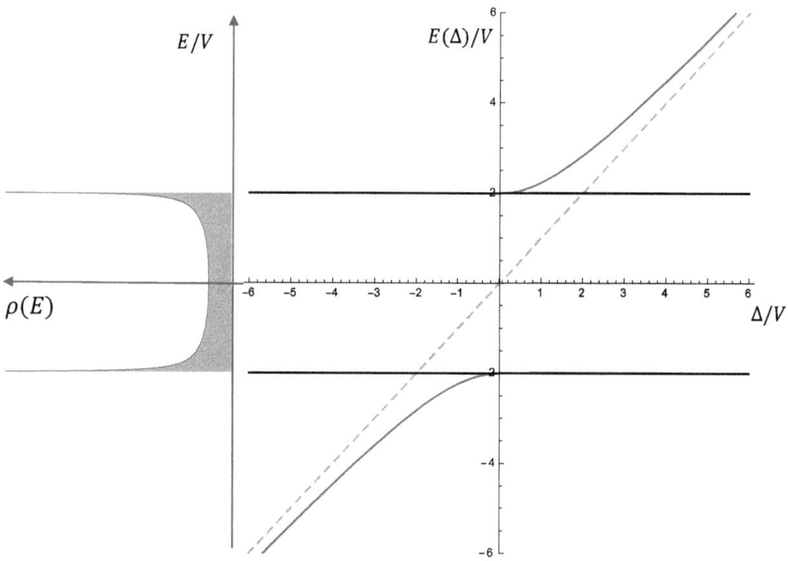

Figure A6.3 The electron eigenenergies in one-dimensional crystal with a point defect as a function of the potential strength Δ/B. The transfer energy is limited to nearest neighbors. The dotted line shows the isolated atomic limit $E(\Delta) = \varepsilon + \Delta$. The density of states $\rho(E)$ is proportional to Im $F_{0ll}(E)$ (Eq. (A6.7b)) which has been shown in Fig. A5.2(a).

For two and three dimensions, the integration can be done numerically, In contrast to one and two dimensions, a localized state appears in three dimensions only when the potential is strong enough: $|\Delta/6V| > 0.34\ldots$. When the potential is very strong $|\Delta| >> V$, the energy of the localized state behaves as $E(\Delta) \sim \varepsilon + \Delta$ for all dimensions.

Hemi elliptic density of states

It would be useful to look at a three-dimensional case where the density of states is given by a following hemi elliptic form with a full width $2B$ (Fig. A6.4)

$$\rho_0(E) = \frac{2}{\pi B^2}\sqrt{B^2 - (E-\varepsilon)^2} \quad |E - \varepsilon| \leq B. \tag{A6.11}$$

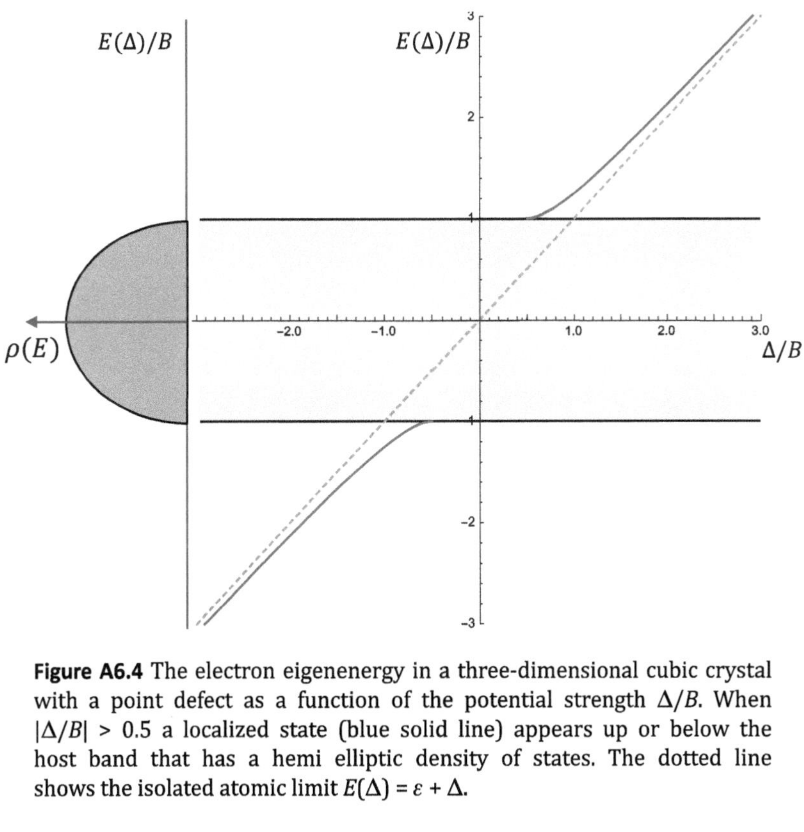

Figure A6.4 The electron eigenenergy in a three-dimensional cubic crystal with a point defect as a function of the potential strength Δ/B. When $|\Delta/B| > 0.5$ a localized state (blue solid line) appears up or below the host band that has a hemi elliptic density of states. The dotted line shows the isolated atomic limit $E(\Delta) = \varepsilon + \Delta$.

The energy dependence near the band edge satisfies the three-dimensional van Hove singularity. The corresponding Green's function is obtained analytically,

$$F_{0\,ll}(z) = \int_{-\infty}^{\infty} \frac{\rho_0(E)}{z-E}\,dE = \frac{2}{B}\left[\frac{z}{B} - \sqrt{\left(\frac{z}{B}\right)^2 - 1}\right]. \qquad \text{(A6.12)}$$

Then, using Eq. (A6.6), the eigenenergy $E(\Delta)$ of a localized state for the defect with a potential depth Δ is given by

$$E(\Delta) = \Delta + \frac{B^2}{4\Delta}. \quad \Delta > 0.5B\text{: repulsive,} \qquad \Delta < -0.5B\text{: attractive} \qquad \text{(A6.13)}$$

absent of localized state, $-0.5B < \Delta < 0.5B$

Application to the adiabatic potential

Here we apply the above results to calculate the adiabatic potentials in a defect system in Section 4.3.3.

An electron moves in a crystal with a point defect and coupled with the lattice distortion. The Hamiltonian we consider is

$$H = |\phi_0\rangle(\Delta - \beta Q_0)\langle\phi_0| + \sum_l |\phi_l\rangle\, \varepsilon\, \langle\phi_l| + \sum_l \sum_{l'} |\phi_l\rangle\, t_{ll'} \langle\phi_{l'}|. \qquad \text{(A6.14)}$$

Here Δ denotes the potential of the defect at the origin and β is the coupling constant of an electron and the lattice Q_0. The other $Q_{n\neq 0}$'s are not considered because they affect little to the lowest electron state. The electronic eigenenergy $\varepsilon(Q_0)$ are obeyed by

$$F_{0\,00}^{-1}(\varepsilon) - (\Delta - \beta Q_0) = 0 \text{ or } F_{0\,00}(\varepsilon) = 1/(\Delta - \beta Q_0). \qquad \text{(A6.15)}$$

with

$$F_{0\,00}(z) \equiv \langle\phi_0|\frac{1}{z - H_0}|\phi_0\rangle = \frac{1}{N}\sum_k \frac{1}{z - \varepsilon_k}.$$

The adiabatic potential is given by

$$U_l(Q_0) = \varepsilon_l\,(Q_0) + \frac{1}{2}CQ_0^2. \qquad l = (\text{f. e.}), (\text{t. e.}) \tag{A6.16}$$

Figure A6.5 shows typical examples of the adiabatic potentials using the three-dimensional Green's function in hemielliptic density of state.

(a) $-0.5B < \Delta < 0$: In the rigid lattice $Q_0 = 0$, no localized state. But as the lattice deforms a localized state split off from the continuum conduction band.

(b) $\Delta < -0.5B$: In the rigid lattice $Q_0 = 0$, there is a localized state. As the lattice deforms the localized state lower its energy. The stable state of S=(t.e.) is approximately given by $\bar{Q}_0 \sim \beta/C$ and the lattice relaxation energy is given by $E_{\text{LR}} \sim \frac{1}{2}C\bar{Q}_0^2 \sim \beta^2/2C$.

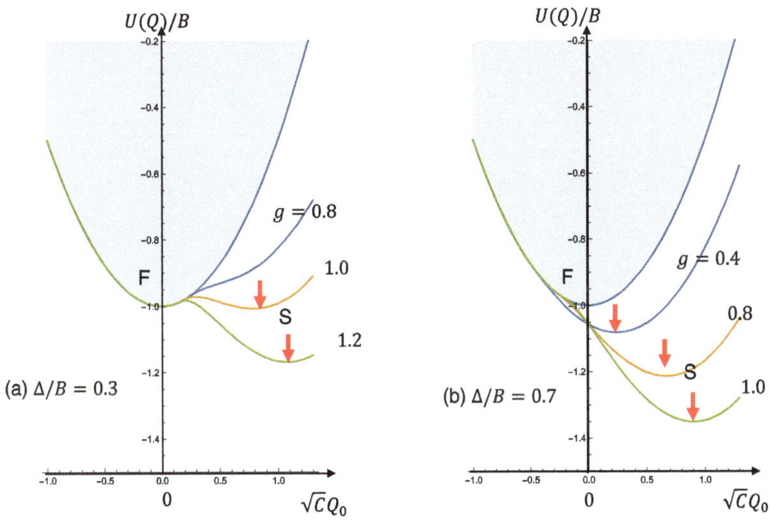

Figure A6.5 Typical examples of the adiabatic potential $U_l(Q_0)$: Δ/B is the relative potential depth, $g = \beta^2/2BC$ is the relative strength of the interaction (see Section 4.3).

Appendix VII

Mid Gap State Coupled with Conduction and Valence Bands

Consider a similar tight binding model in Appendix IV and extend it to cover both the conduction and valence bands. (Fig. A7.1). Hamiltonian of an electron is expressed as

$$H_e = \sum_{n} |\phi_n^c\rangle \varepsilon_n^c \langle \phi_n^c| + \sum_{n\neq 0}\sum_{m\neq 0} |\phi_n^c\rangle\, t_{n,m}^c \langle \phi_m^c|$$

$$+ \sum_{n} |\phi_n^v\rangle \varepsilon_n^v \langle \phi_n^v| + \sum_{n\neq 0}\sum_{m\neq 0} |\phi_n^v\rangle\, t_{n,m}^v \langle \phi_m^v|$$

$$+ \sum_{n\neq 0} \left(|\phi_n^c\rangle t_{n,0}^{cv} \langle \phi_0^v| + |\phi_0^v\rangle t_{0,n}^{vc} \langle \phi_n^c| + |\phi_n^v\rangle t_{n,0}^{vc} \langle \phi_0^c| + |\phi_0^c\rangle t_{0,n}^{cv} \langle \phi_n^v| \right). \tag{A7.1}$$

Here ϕ_n^σ denotes the Wannier orbital in σ = c, v band at the site n, whose energy has a form

$$\varepsilon_n^c = \begin{cases} \varepsilon^c + \Delta_c & n = 0 \\ \varepsilon^c & n \neq 0 \end{cases}, \quad \varepsilon_n^v = \begin{cases} \varepsilon^v + \Delta_v & n = 0 \\ \varepsilon^v & n \neq 0 \end{cases} \tag{A7.2}$$

When (a) $\Delta_c, \Delta_v < 0$, an electron in the conduction band is attracted by the defect at the origin (n = 0). When (b) $\Delta_c, \Delta_v > 0$, an electron in the valence band is repulsed by the defect at the origin; hence a hole, if it exists in the valence band, is attracted. The transfer energy to/from the defect state is modulated by a factor ξ^c (ξ^v, ξ^{vc}, ξ^{cv}) of the relevant transfer in the host.

$$t_{n,0}^c = \xi^c\, t_{n+m,m}^c, \qquad t_{n,0}^{vc} = \xi^{vc}\, t_{n+m,m}^v, \tag{A7.3a}$$

$$t_{n,0}^v = \xi^v\, t_{n+m,m}^v, \qquad t_{n,0}^{cv} = \xi^{cv}\, t_{n+m,m}^c. \tag{A7.3b}$$

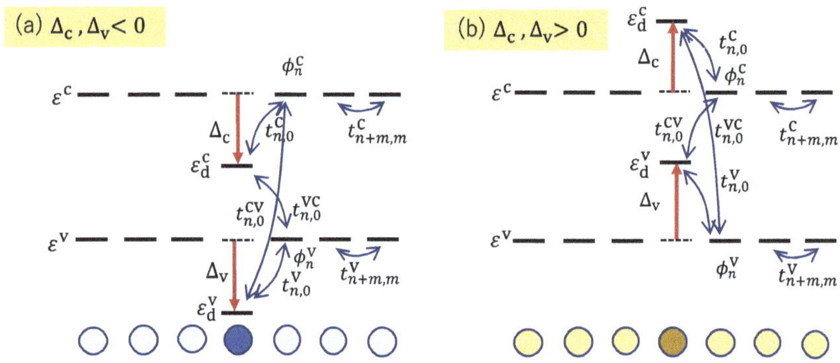

Figure A7.1 A tight binding model, where electron states are mixed via a defect with potential difference and modulation in transfer energies.

A procedure to obtain general solutions for the eigenenergy of the Hamiltonian (A7.1) is given elsewhere [A7.1]. In particular, for a typical situation where one of the Wannier states at the defect, for example ϕ_0^{v} is isolated (Δ_{v} << 0 and/or $\xi^{\mathrm{v}} = \xi^{\mathrm{cv}} = 0$), the solutions are given by

$$\varepsilon^{\mathrm{c}} + \Delta_{\mathrm{c}} = (\xi^{\mathrm{c}})^2 \left[\frac{1}{F_{00}^{\mathrm{c}}(E)} - (E - \varepsilon^{\mathrm{c}})\right] + (\xi^{\mathrm{vc}})^2 \left[\frac{1}{F_{00}^{\mathrm{v}}(E)} - (E - \varepsilon^{\mathrm{v}})\right] + E, \tag{A7.4a}$$

$$\varepsilon^{\mathrm{v}} + \Delta_{\mathrm{v}} = E. \tag{A7.4b}$$

Here $F_{00}^{\mathrm{c}}(E)$ and $F_{00}^{\mathrm{v}}(E)$ is the diagonal part of the Green's function in the host band defined by

$$F_{00}^{\mathrm{c}}(E) = \frac{1}{N}\sum_{k} \frac{1}{E - \varepsilon^{\mathrm{c}} - t_k^{\mathrm{c}}}, \quad F_{00}^{\mathrm{v}}(E) = \frac{1}{N}\sum_{k} \frac{1}{E - \varepsilon^{\mathrm{v}} - t_k^{\mathrm{v}}}, \tag{A7.5}$$

with the respective energy spectrum

$$t_k^{\mathrm{c}} = \frac{1}{N}\sum_{n}\sum_{m} \exp[i\boldsymbol{k}(\boldsymbol{R}_n - \boldsymbol{R}_m)]\, t_{n,m}^{\mathrm{c}}, \tag{A7.6a}$$

$$t_k^{\mathrm{v}} = \frac{1}{N}\sum_{n}\sum_{m} \exp[i\boldsymbol{k}(\boldsymbol{R}_n - \boldsymbol{R}_m)]\, t_{n,m}^{\mathrm{v}}. \tag{A7.6b}$$

Furthermore, when $\xi^{vc} = 0$ and $\xi^{c} = 1$, the solutions of Eq. (A7.4a) is reduced to those of Eq. (A6.6) for a single band.

Figure A7.2 shows results on a typical case $\xi^{v} = \xi^{vc} = 0$ where the energy spectrum of the host valence and conduction bands both have a hemi-elliptic density of states.

$$\rho^{\sigma}(E) = \frac{2}{\pi B_{\sigma}^{2}}\sqrt{B_{\sigma}^{2} - (E - \varepsilon^{\sigma})^{2}}, \quad |E - \varepsilon^{\sigma}| \leq B_{\sigma}^{2}. \qquad \sigma = \mathrm{c}, \mathrm{v} \tag{A7.7}$$

The black solid line indicates the energy level of purely isolated defect state ϕ_{0}^{c}: $\xi^{c} = \xi^{vc} = 0$. As ξ^{c} increases, the mixing of ϕ_{0}^{c} and near $\phi_{n}^{c'}$'s shifts the defect level downward in the band gap. As ξ^{vc} increases, the mixing of ϕ_{0}^{c} and near $\phi_{n}^{v'}$'s shifts the defect level upward in the band gap.

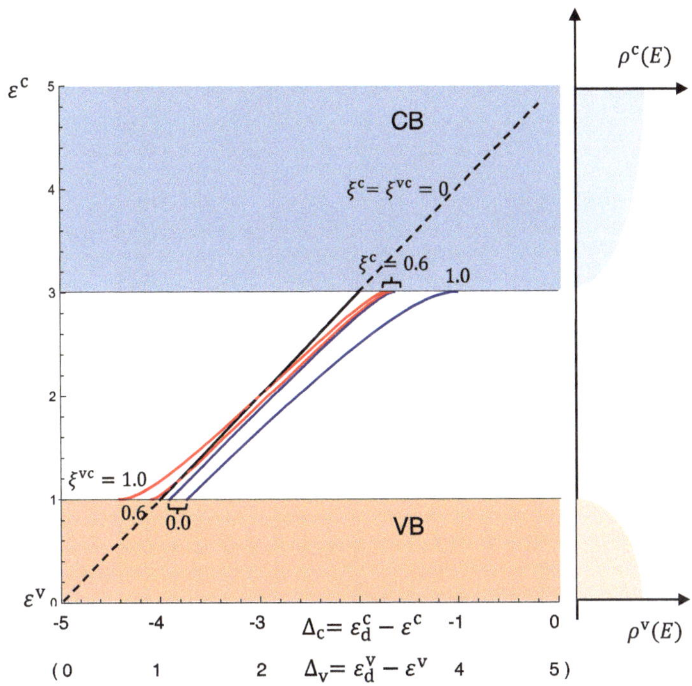

Figure A7.2 Typical solutions of eigenenergies of Hamiltonian (A7.1) as a function of potential difference $\Delta_{c} < 0$. Parameters used are: $\varepsilon^{v} = 0$, $\varepsilon^{c} = 5$, $B_{v} = 1$, $B_{v} = 2$, and $\xi^{c} = \xi^{vc} = 0$. This result can be applied for the dependence of potential difference $\Delta_{v}(=5 - \Delta_{c}) > 0$ in the valence band with $\xi^{v} = \xi^{cv} = 0$.

Reference

[A7.1] Shinozuka, Y., *J. Phys. Soc. Jpn.*, **51** (1982) 2852.

Appendix VIII

Anderson's Negative-U

In 1975, Anderson showed the mechanism of negative-U, which indicates a possibility that the effective interaction between two electrons become attractive in spite of the Coulomb repulsion between them [A8.1]. We consider here a primitive case: two localized electrons interact with the lattice vibrations in addition to the Coulomb repulsion between them. The Hamiltonian we consider is (Fig. A8.1)

$$H = \sum_{i=1,2} \left\{ \sum_{\sigma=\uparrow,\downarrow} \varepsilon \, n_{i\sigma} + U n_{i\uparrow} n_{i\downarrow} - \beta (n_{i\uparrow} + n_{i\downarrow}) Q_i + \frac{C}{2} Q_i^2 \right\}. \quad \text{(A8.1)}$$

There are two sites i = 1, 2: $n_{i\sigma} \equiv a_{i\sigma}^* a_{i\sigma}$ is the electron number operator with spin $\sigma = \uparrow, \downarrow$ at the i-th site. $U(>0)$ denotes the Coulomb repulsion energy when two electrons are located at the same site: on-site U. Q_i is the lattice distortion at i-th site, β the coupling constant and C the elastic constant. Electron transfer energy is omitted.

When the system contains two electrons, the eigenstates are (a) single-occupancy at both sites and (b) double-occupancy at one of two sites. The corresponding adiabatic potentials are, for example,

$n_{1\sigma} n_{2\sigma'} |0\rangle$

$$E_{\text{single}}(Q_1, Q_2) = 2\varepsilon - \beta Q_1 - \beta Q_2 + \frac{C}{2} Q_1^2 + \frac{C}{2} Q_2^2, \quad \text{(A8.2a)}$$

$n_{1\uparrow} n_{1\downarrow} |0\rangle$

$$E_{\text{double}}(Q_1, Q_2) = 2\varepsilon + U - 2\beta Q_1 + \frac{C}{2} Q_1^2 + \frac{C}{2} Q_2^2. \quad \text{(A8.2b)}$$

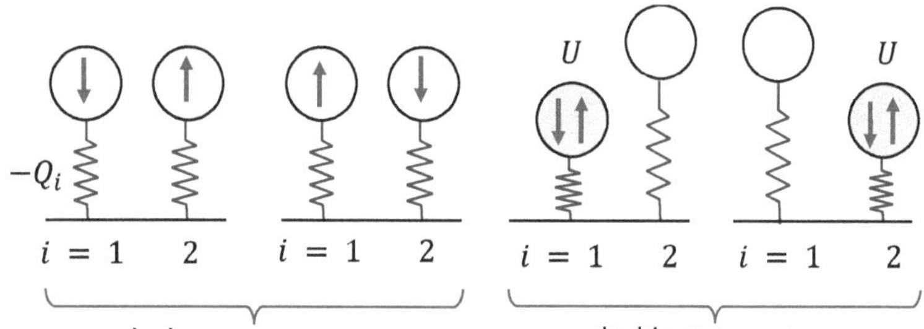

Figure A8.1 Two-site two-electron model.

The respective minimums are given by

$$E_{\text{single}}(\beta/C, \beta/C) = 2\varepsilon - \beta^2/C, \quad \text{(A8.3a)}$$

$$E_{\text{double}}(2\beta/C, 0) = 2\varepsilon + U - 2\beta^2/C. \quad \text{(A8.3b)}$$

Then, if $\beta^2/C > U$, the ground state is two electrons localized at one site in spite of the Coulomb repulsion U between them. Thus, the effective repulsion energy $U - \beta^2/C$ is negative, which is called the Anderson negative U.

Related phenomena are found in various systems: Bipolaron [A8.2], Magnetic deep-impurity in semiconductor [A8.3]. If we further take account of electron transfer between different sites, the character of the ground state and excited states will have further variety: see the *T-U-S* model introduced by Toyozawa [A8.4].

References

[A8.1] Anderson, P. W., *Phys. Rev. Lett.* **34** (1975) 953.

[A8.2] Alexandrov, A. S. and Mott, N. F., *Polarons and Bipolarons*, World Scientific, Singapore, 1995.

[A8.3] Matsuda, T., and Ohata, N., *Solid State Commun.* **39** (1981) 678.

[A8.4] Toyozawa, Y., *J. Phys Soc. Jpn.* **51** (1981) 1861.

Index